BIOLOGY TOPIC TEXT E: ECOLOGY AND BIOGEOGRAPHY PRIMER

HAROLD M. ICKES II, M.D.

Executive Editor: Rhett C. Smith, Ph.D.

Ideal Quill Publishing Group, LLC

idealquill.com

Printed in the United States of America

10 9 8 7 6 5 4 3 2 1

ISBN 978-0615834863

Biology Topic Text E: Ecology and Biogeography Primer

by Harold M. Ickes II, M.D.

Edited by Rhett C. Smith, Ph.D.

Table of Contents

Unit 1

Ecology I: Population Size

In this section we cover:

1) Basic ecological definitions
2) Population growth models
3) Factors affecting population growth

Populations, Communities, Ecosystems and the Biosphere

Until now, we've been up to our knickers in the specifics of chemicals, cells, inheritance and finally evolution. Now we make a quick transition to how groups of organisms tend to interact with each other. This area of study is broadly known as **ecology**, and it includes many factors found in physical reality and how these factors influence the survival of a species.

Before we get all bothered over the details, we need to lay down a few definitions. First is the definition of **population**. We saw this word before in Topic Text A, but that feels like years ago, so let's rehash it: a population is defined as a localized number of individuals belonging to the same species. So the individuals of the species of fungus left in my fridge by my roommate represent one population of that species of fungus. The same species found in *your* fridge (left there presumably by your roommate, of course) is a different population of that fungus.

Next on the list of definitions is the **community**. The community is the group of interacting species within a given area. So, the fungus, the bacteria and whatever else he left in my fridge all comprise a community of different species, all interacting with each other. Your local forest is a great example of a community of living organisms – a large variety of different species interacting with each other.

Getting one step deeper into the definitions, we now encounter the **ecosystem**, which is the community plus the physical nonliving structures in and around the community. So, the insides of my fridge – the shelves, the lining, the light he broke, and so forth – are sadly a part of the ecosystem growing in my fridge. I would love to add cleansing materials to that ecosystem, but I honestly fear from my life; I'd rather the fungus remain out of *my* immediate ecosystem, after all.

The example of my fridge is rather small-scale. Usually all these terms are used to define much larger examples. A forest contains many populations, since there are many different species. The forest is a community, since there are many populations within it. The forest also represents an ecosystem, because it typically contains nonliving elements, such as rocks, water, air and dirt, among other things.

To wrap up all these definitions, all the ecosystems on the planet are contained within the **biosphere**. The biosphere encompasses the relatively thin shell of the Earth's surface. Most life on earth is contained there. The deeper you go in the Earth, the less life you encounter. Likewise, above the clouds the incidence of life drops off as well. Since there is only one Earth,[1] it means there is only one biosphere. The biosphere is the highest level of organization in biology, and it is virtually a closed system in terms of materials. That means that atoms rarely escape from it into outer space. The only thing that enters and leaves the biosphere is energy: the Sun's energy

---

[1] I'm sure this is a very reassuring thought to all the aliens out there in the cosmos, each holding us at arm's length. They probably complain that their roommate left us here to ruin a perfectly good solar system.

comes in and heat radiates out. All other materials cycle around over and over again.

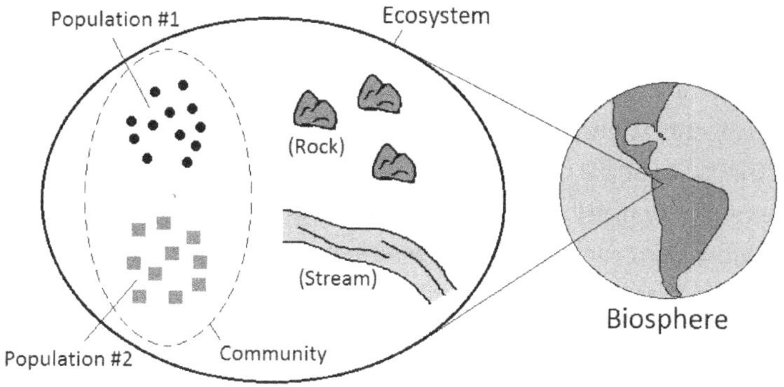

Population Shape and Size

Now that we know that a population is a localized collection of individuals that are all of the same species, let's dig a little deeper into the subject. Specifically, let's look at how populations appear and how they grow or shrink in size. We'll have to be somewhat general and do a bit of hand-waving . . . after all, animals and plants come in a huge variety of shapes, sizes and living styles.

One particular aspect of populations that scientists are often concerned about is how spread out or how "distributed" a population is. Are the individuals living closely together? Are they spread out and solitary? Is the distribution uniform or random? The answers vary, but how individuals are distributed often falls into one of three categories, shown in the image that follows.

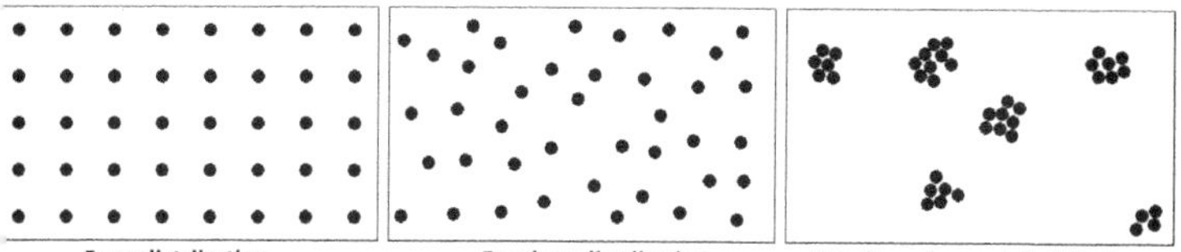

Even distribution | Random distribution | Random clumped distribution

Of the distribution types shown, the random distribution and randomly clumped distributions seem to be more common than evenly spread-out populations. Evenly spread-out populations are unlikely, because the physical environment is rarely uniform enough to allow animals to spread out evenly. As such, nonsocial species spread out wherever their environment allows, creating a random distribution. Social organisms usually clump together,[2] and the clumps are randomly distributed based on where the environment allows. In addition, clumping tends to be where major resources can be found, or where there is a minimum of environmental hazards, and so forth. Consider human cities, for example. Our cities tend to be associated with oceans or major water ways, primarily because water constitutes a major resource and means of mass transit (especially in the old days).

On the other hand, clumping isn't always an indication of social behavior. Take the example of water again: if there's a water hole in the desert or other arid place, you'll often see a bunch of organisms clumped around it, regardless of how social they would be otherwise.

Another strong motivator for clumping is reproduction Asexual reproduction results in offspring that are nearby, such as might occur in many fungi. Similarly, sexual reproduction results in clumping because of the need to find a mate. In the latter case, some nonsocial populations are clumped only during mating seasons, after which they return to being spread out.

---

[2] ... with exception to grumpy bio book writers, but I suppose that falls under nonsocial.

Here a writer in his natural habitat prepares to defend his watering hole against a deer.

Other basic aspects of populations that scientists are often concerned about are population size and total biomass of the species[3]. Let's focus on the population size, or simply: the number of individuals in the population. How do we measure this? Consider plants first. They are usually not particularly hard to count, and to accomplish the task we often use a technique called the **quadrat** method. This entails dividing up the region into squares and then counting the number of organisms within a random sampling of squares. Once you have counted a sufficient number of squares, you can assume the other squares are similar. So, if 20 out of 100 squares each have an average

---

[3] Usually directly related to the number of organisms in a population, the total biomass just represents how much mass the population has. This piece of information can sometimes tell you about the nutritional state of the population. If there are a lot of individuals in the population, but the total mass is surprisingly low, it suggests the population is starving. At other times, it becomes too difficult to distinguish 'individuals' of the species, so the total biomass serves as a more efficient measurement to estimate population size.

of 15 plants in them, then we can assume that the total number of plants in the region is roughly 100*15 = 1500.

Now, in theory this method works for animals as well, especially those that are relatively slow or those that do not migrate, like snails or spiders and several of my relatives . . . but what about the fast moving animals? How in the world do we count *them*? They are constantly moving this way and that, migrating and scattering, especially the moment *we* show up.[4]

Well, one method, called **marked capture** (or mark and recapture[5]) involves the initial capture and tagging of a bunch of animals in the wild. At a later time, animals in the region are captured again, and the percentage of those wearing the tag gives an indication of how many animals are in the region. For example, suppose we captured and marked 100 fish within a lake. Two days later, we captured 100 more and only 10% of those were tagged. From that, we can guess that one out of ten fish in the lake are tagged. Since there are 100 tagged fish in the lake, there must be around 1000 fish total. The disadvantage of this method is that it does not take into account animals born, animals that die, or animals that migrate in and out of the area.[6] You can use this method on all kinds of critters, from frogs, to deer, to rabbits, to the children in your neighborhood.[7] Using this technique, I've managed to determine that there are roughly 2000 children in my neighborhood, and with more advanced techniques I've determined that most of them are clustered in the next apartment

---

[4] ... which proves they are smarter than we give them credit for.

[5] There are a bunch of variants to this name. Tom-AY-to, Tom-AH-toh, I say.

[6] In this particular case, animal death is probably the most relevant, assuming no fish were born in the short time between capture, and also assuming none packed their little fish bags and migrated to another lake. I try to keep an open mind.

[7] Relax. You can tag them harmlessly by giving them goofy hats. What were you thinking?

Will the one with the blond hair *please* give the one with the brown hair his toy back? All the humanity!

Now we have a basic idea of how we measure the size of a population, but why would we want to know this, anyway? Well, by knowing the size of the population and observing how the size changes over time, we can get an idea of how populations are affected by their environment – by other species, by physical forces, etc. If species A increases in number (for whatever reason) leading to a decrease in species B and C, it suggests that there may be an association between species A and species B and C. This can spur more research that leads to more discovery. In effect, measuring population size is a way for us to trace *possible* causes and effects in an ecosystem, with suitable hand-waving, of course. Think of measuring population size as being as fundamental to an ecologist as a ruler is to a builder . . . at least that's what the tv cartoons next door suggest. It's a lesson so vital that they play it around the clock.

Before we knock ourselves out by the little bits of math that are about to follow (ack!), let's jump to what we know about common population growths. They usually start out increasing in size and then at some point they stop growing as fast. Eventually, they stop increasing in size altogether. This kind of model is called logistic growth. It over-simplifies reality, but does well for an approximation. It looks roughly like the following image.

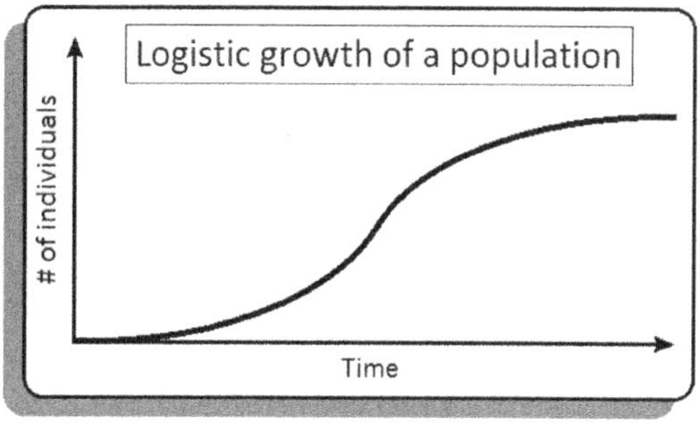

Let's see if we can come up with a math equation to match this. Why? Well, having an equation with lots of fancy letters in it often helps us to target factors in reality that are influencing the trend we are observing. For example, suppose we know that the trend follows an equation that has two variables in it. This suggests that there could be two primary factors in reality we should look for. Sometimes the act of trying to derive an equation based on our assumptions teaches us about the errors of our assumptions. Either way, let's give it a shot and see what happens.

First, let's imagine a situation in which everything is ideal for growth. For example, what do you suppose would happen if, say, each organism gave rise to two offspring, and all the offspring survived to reproduce as well? Further, what if none ever died or moved away? The first individual would give rise to two more. These two would give rise to four. Four begets eight, which begets sixteen, and so forth. Before long, the growth would turn into a ridiculous exponential explosion covering the earth in whatever organism we are talking about, and knowing my luck they would all move next door to me.

Now, if a population really did grow this way, it would follow an equation something like this:

$$N_{new} = 2N_{current}$$

This means: the number of new individuals, N_{new}, is twice as much as the current number of individuals, $N_{current}$. This form of the equation is not incredibly useful though, because it only churns out numbers one generation at a time. To do this properly, we'll need to describe the equation using some calculus and differential equations . . . which is WAY outside the scope of this book, thankfully. So, for the moment, you'll need to trust me that the above equation yields the next one after dropping it into a mathemagic box and transforming it: *POOF*

$$\Delta N/\Delta t = rN$$

Don't panic. The little triangles just mean "change in," so the above equation says the "change in the number of individuals over a period of time equals rN," where r is a number. In this case, r is just 2. The graph for such a growth is as follows.

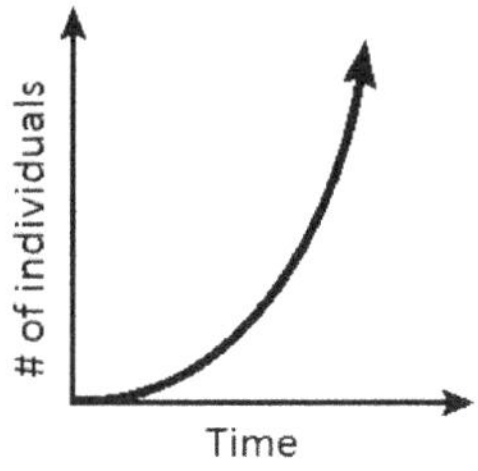

The graph accurately represents how such an ideal population would grow over time. Unfortunately, this does not match our original graph, so we'll need to dig a little deeper. Most of the time, the environment is usually full of predators, parasites, harsh weather, illnesses, diseases, mother-in-laws, final exams, taxmen and a whole slew of other miserable things. All of these things can bring death,[8] so instead of considering only the births, let's also consider the deaths as well. We replace "r" with "b-d," or "births minus deaths":

$$\Delta N/\Delta t = (b - d)N$$

So, if there are 2 births per parent and each parent died, then b-d would 2-1 = 1, causing $\Delta N/\Delta t$ to be equal to N. That means the population would increase by a size equal to whatever generation it was, designated as N. Like the first situation, this also leads to exponential growth, but much less severely. What about if the rate of births equals the rate of deaths? In this case b-d = 0, and $\Delta N/\Delta t = 0$, meaning the population does not change size. It has reached equilibrium. Finally, what if the deaths outnumber the births? If this is so, then b-d < 0, and the number of individuals decreases per generation.

Ok, maybe final exams don't bring death, but they certainly feel like it. Surviving the board exams was nothing short of a Darwinian marvel for me.

To summarize:

Births exceed deaths	-> population increases
Births equal deaths	-> population stays same size
Deaths exceed births	-> population decreases in size

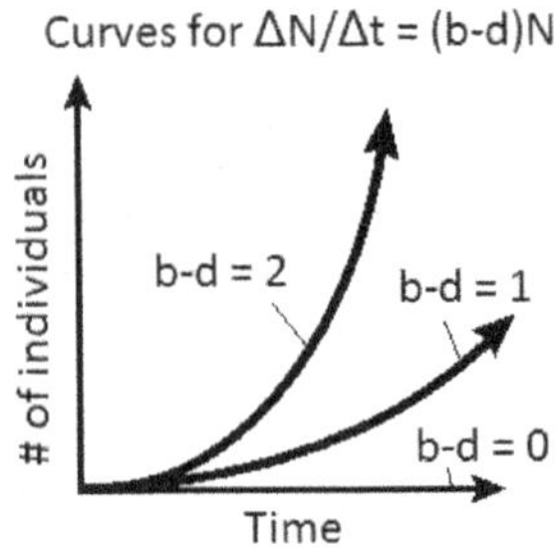

I've included the nifty little graph to the left to show how the population size changes over time for several different values of b-d. Notice how the rate of growth gets smaller and smaller as b-d gets closer to zero.

Still, none of these reflect our initial graph! Even after taking into account deaths, we can't seem to achieve our goal. So, instead of horsing around looking for the equation, let's do what scientists often do and borrow an equation for mathematicians and see how far off we were.[9] The equation looks like this:

$$\Delta N/\Delta t = r_{max}(K-N)/K \ * N$$

If you graph it, you get our original graph. I've posted it again below with a few extra details.

[9] I snuck into the math department last night and found the equation under an old coffee cup. I doubt they'll miss it. This particular equation is often called the Verhulst equation, after Pierre-François Verhulst.

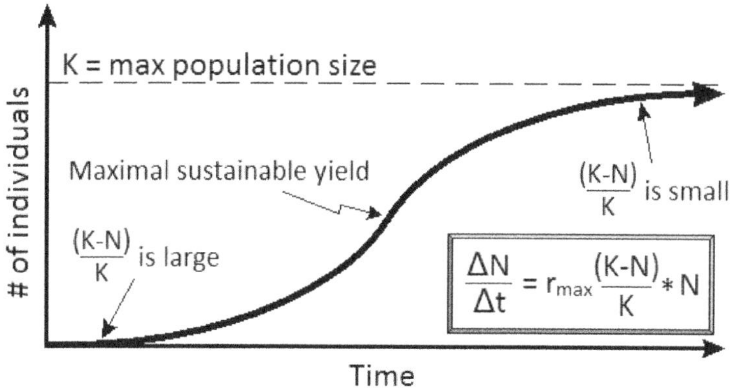

I know what you are thinking: what is the "r_{max}(K-N)/K" gobble-dee-goop? Since a hazy film may be forming over your eyes, I'll try to simplify this mess. All of these equations follow the pattern $\Delta N/\Delta t = xN$. In our first attempt, we tried using just r where x is. That led to an exponential growth, which doesn't match the target graph. We then tried b-d in place of x, to reflect the births *and* deaths. That *still* led to exponent growth (positive and negative) or to zero growth. Now we are trying r_{max}(K-N)/K in place of x. Here, the r_{max} is the maximum rate of growth. It is called the **intrinsic rate of increase** and it represents the rate of growth under ideal conditions. Next, the K refers to the maximum number of individuals that the population appears to be able to hold, designated by the horizontal dotted line. N is the number of individuals in the population at any given period of time. This equation works. Why? It succeeds because *the value of (K-N)/N changes with time*. Specifically, as the number of individuals, N, approaches the maximum, K, the top part of the (K-N)/K approaches zero. This forces $\Delta N/\Delta t$ to approach zero as well, which means that the population growth slows down as it approaches the top of the graph.

Since the equation works because (K-N)/N changes with time, this suggests that our (b-d) attempt would have worked if b and d changed with time. That is the first lesson we get from our math challenge: the rate of births and deaths both change in time. The second lesson is that they seem to be related to the maximum number of individuals the environment can hold. The closer the population gets toward this maximum size, the greater the number of deaths and/or smaller the rate of births. When the population is small, births clearly outnumber the deaths. When the population is near its maximum, the number of deaths becomes roughly equal to the number of births. Smack dab in the middle of the graph is a very brief point in which the rate of increase is at its highest (a point called the **maximal sustainable yield**).

When all is said and done, this analysis tells us that the *rate of births and deaths are related to the population size.* That leads us to look for factors that relate population size to birth and death rates.

Consequences of High Density Populations

How might population size affect birth and death rates? Well, as the number of individuals within a population increases, it slowly becomes harder and harder for each individual to survive. The first – and probably most obvious – factor is competition for resources. Think of it this way: if your school cafeteria was your sole source of nutrients and it began running out of food, you'd have to fight your fellow students for the last scraps.[10] You would starve if you couldn't compete. Nature works in a similar way. The limit on natural resources acts as a built-in mechanism to keep the number of organisms in check. It usually works by killing off the weaker members of the group because these individuals are the least able to compete.

[10] Well, that's perhaps a bad example. From the school cafeterias I've been to, the less food they had the happier everyone seemed to be.

The second reason that deaths increase with population size is that parasitism and disease increase in incidence. The more organisms that exist in a population, the more targets there will be for bacteria and viruses. After all, the population of organisms *represents the primary resource* for bacteria and viruses! If the population increases, then the bacteria and viruses have an over-abundance of resources at their disposal, which leads to an increase in growth of these pathogens. Once again, all this parasitism and disease usually leads to the death of weaker members of the population.

Next is predation. As the population gets larger, it experiences more death due to predators (assuming they are around). This last factor can be quite deceiving, however, because the predators and prey in a stable ecosystem usually exist in proportion to each other.[11] For example, for every 100 deer, you might have, say, 1 wolf. For every 1000 citizens, you'll have one lawyer. The more the predator consumes, the larger the size of the population needs to be. That's probably why there are only a handful of presidential candidates per 100 million people . . . but that's just a guess. I threw out my TV years ago. I get most of my news via Pony Express.

Anyway, if the population got ten times as big, the predator population will also get ten times as big, assuming no other complications. This in and of itself will not impose a limit on the population's growth. In fact, strength is often found in numbers (assuming the population is social). The more members of a group, the more eyes there will be to watch for danger. It's a lot easier to sit and eat when you have people guarding your borders. When predation does occur, it usually takes the weakest members of the

[11] Or more correctly, they tend to fluctuate around a proportion in a cyclic fashion. If the prey increases in number, the predators will slowly increase in number, after an initial delay. If the prey drops in number, the excess predators will linger for a time then drop off as well.

group. It's like the old adage: You don't need to run the fastest; you only need to run faster than the slowest person in the group. That's evolution for you - survival of the fit-enough, and most of us have family members that are living proof.

While disease, parasitism and predation certainly play their part, the absolute limit on the size of the population ultimately falls to limits placed by the physical environment, usually from the availability of resources and space. In fact, when it comes to these things, organisms often have to compete against more than just other members of its own species. Quite often, one species will have to compete with another, creating a situation referred to as **interspecies competition**.

The degree to which two species will compete with each other for survival generally comes down to how similar their **niches** are. What is a niche? Well, put simply, the niche is a catch-all term that basically describes all the aspects of their living strategy. It encompasses things like: what they eat, where they nest, where they mate, the time of day they are active, their resistance to environmental change, the pub they frequent, and so forth. If two species use the same nesting grounds at the same time, say, then they will tend to butt heads a lot. If they eat the same things, then they will tend to fight a lot, especially after frequenting their aforementioned pub. Anyway, if two species are too similar in their niche, one of four possible outcomes usually results:

1) One species will eventually go extinct.
2) The two competing species will gradually become segregated to two separate geographic locations. In effect, they establish a boundary. On one side, species A will have defeated species B. On the other side, the opposite has occurred. This kind of phenomenon is often

referred to as **allopatriation**[12]. They start as **sympatric**[13] and become **allopatric**.

3) One species will give the other a sound defeat, but will experience periodic mass death by regular environmental forces. This temporarily allows the weaker species to rebuild its numbers, setting the stage for the next round. This kind of battle tends to be a temporary and unstable condition. Eventually, one species will usually win, resulting in (1) or (2).
4) The two competing species will gradually acquire changes in lifestyle that minimizes the amount of niche overlap. This may occur through change in behavior or through evolution. For example, one might switch the kind of animal it hunts, or it may evolve a change in beak size so that it can eat a different kind of seeds. If they nest in the same area, the two species might evolve a change in nesting sites. Sometimes, they can even evolve **mutualism**, where the two species begin working together to help each other survive. Insert happy fuzzy feeling here.

All of these outcomes alleviate the strain of competition between two species, and these patterns are strong enough that it has led to a hard-and-fast rule in biology: *no two species can occupy the same niche*. Of these outcomes, mutualism is perhaps the most fascinating and deserves a closer look.

Mutualism

---

12 This means, roughly, "different nations."

13 "Occupying the same nation"

Mutualism is a phenomenon in which two species have evolved to coordinate their niches so that both benefit. A classic example of this includes mitochondria and chloroplasts living inside eukaryotic cells. Believed to once have been free-living, the mitochondria and chloroplasts were probably parasites on early eukaryotic cells, invading one cell and then moving on to the next. At some point, both parasite and prey evolved a genetic relationship in which the mitochondria and chloroplasts began assisting the eukaryotic cell, which in turn gave the mitochondria and chloroplast a home. Other examples (usually covered in the second semester of biology) can be found in the study of botany, such as a number of mutualistic relationships involving plant roots. These include the nitrogen-fixing bacteria and fungal mycorrhizae. These latter two organisms live in and/or around plant roots, benefiting the plant while receiving benefits from the plant in return. Presumably the start of this relationship was not lovey-dovey though, no doubt taking millions of years of cellular warfare before a truce was drawn.

Either way, let's get back to the point. For the reasons given above, high density populations are often disadvantageous to the species in question. Organisms in overcrowded areas sometimes leave the population.[14] This emigration behavior is very risky and very energetically costly though. As such, migration is usually only done

[14] . . . excluding bullheaded individuals, such as yours truly.

when it's *more costly to stay than to leave*. Overcrowding can also spark internal regulation of reproduction as well, where mating pairs will sometimes forgo reproducing because resources have become too sparse. For example, long term restriction of calories (i.e. starvation) is well known to cause a lot of species to abstain from reproducing. The reason for this is quite logical: if there aren't enough calories around to sustain the current generation, then newborns will not survive and the entire process will only spread out the calories even more. You *and* your children could die, so the best bet is not to mate at all until the conditions get better.

Advantages of High Density Populations

Despite all this business of death and destruction in high density populations, there are definite advantages to having large numbers of cohorts. In plants, for example, an increase in density often leads to a reduction in wind. In addition, dense forests are usually better able to minimize temperature and moisture fluctuations, to the benefit of all within the forest. For social animals, as we mentioned above, an increase in the members of the society can lead to more complex interactions and greater cooperation. Not only are more members of the society looking out for common enemies, but more are able to take on specialized roles within the population. Humans are an excellent example of this concept, and despite the frequency of doomsdayers who point to the logistic graph and predict our extinction by overcrowding, I quickly point to our technological advantages. While we have a keen knack for mucking up environments, we have just as keen a possibility of getting off this rock to muck up other planets as well. The future is brighter than ever!

Well, that about covers our initial look at ecology. Now that we've discussed the basics of populations, population size, and the effects of overcrowding, let's move on to examine the effects of changing environments. See you up ahead.

Summary

In this chapter, we covered the basics of ecology:

- Broadly, ecology is the study of how organisms interact with their environment and with each other.
- A population is a group of individuals within the same species living in a geographic region.
- A community is a group of populations (of different species) living in a geographic region.
- An ecosystem is a community plus the physical objects in it, such as rocks, streams, wind, etc.
- The biosphere is the shell around the Earth that contains all forms of life. It is materially a closed system, meaning that practically nothing escapes it into outer space. Energy enters in the form of the Sun's light and leaves in the form of heat that radiates out into space.
- Populations can have even distribution, random distribution, or randomly clumped distribution. Of these, the random distribution and randomly clumped distributions are the most common. Even distribution is rare because the physical environment is usually not uniform enough to allow this form of distribution. Random clumping is generally caused by the location of physical resources. It is often an indication of social behavior, but not always.
- Measuring population size is important to ecologists because it helps them trace possible cause and effect through communities of organisms. It is an essential measurement in their field. There are two major ways that populations are measured:
    - Quadrat method – the region is divided into squares and organisms within a random sampling of squares are counted. The rest of the squares are assumed to

hold a similar number of individuals. The total number of organisms is assumed to be the average number of individuals in the sampled squares times the total number of squares.

- Marked capture – a subset of individuals within a population are captured, marked and released. Shortly later, another sampling of individuals is captured, and the number of individuals in the second set that are marked gives an indication of how many individuals are in the entire population.

- Population growth often follows a logistic model, described by the equation:
  - $\Delta N/\Delta t = r_{max}(K\text{-}N)/K \; * \; N$

    Where $\Delta N/\Delta t$ is the change in the population's size, r_{max} is the rate under ideal conditions, K is the maximum number of individuals the environment can support, and N is the population size at any given point in time. This equation tells us that the growth of the population depends on how close it gets to the maximum size it can have. It also tells us that the rate of births and rate of deaths change over time. Specifically, the rate of deaths tends to increase as the population nears overcrowding.
- Overcrowding leads to increased death due to several primary factors:
  - Limit of natural resource
  - Parasitism and disease
  - Predators

    Of these factors, the first is usually the main determinant of the maximum size a population can be.

- The niche is a catch-all term that describes the lifestyle of the species in question. It includes what the species eats, where it nests, the method of mating, the times of year it is active and so forth. The total number of variables encompassed by the niche is enormous and generally unknowable.
- Overcrowding usually results in the death of the weakest members, but it also can spark competition between species as well. This is called interspecies competition. Two species having highly overlapping niches will result in fierce competition. The result of this competition usually falls into 1 of 4 categories:
    - One species will go extinct.
    - The two species will become allopatric.
    - One species will nearly win, but will be periodically weakened by external forces, during which time the losing species will regain ground. The process will then repeat.
    - The niches of the two species will change to minimize competition. This occurs by elective changes in behavior or through evolution. Sometimes the two species will evolve a mutualistic behavior that benefits both species.
- Mutualism is where two species evolve to work closely together, to the benefit of both species.
- Overcrowding can sometimes lead to emigration, where it becomes less costly to move away than it is to stay. Overcrowding can also lead to internal regulation of breeding. Caloric restriction, presumably through competition for food, is one example in which many species reduce their breeding to compensate for the lack of resources.

- The primary benefit of high density populations is through the potential to work together through social interaction. In plants, high density can foster favorable microenvironments that reduce wind, temperature and moisture fluctuations.

Unit 2

Ecology II: Ecological Change

In this section we cover:

1) Ecosystem change and reproductive strategy
2) Survivorship curves
3) Succession
4) Ecosystem Stability

The Changing Environment vs. Reproductive Strategy

In the previous discussions, we attempted to derive an equation to show how population size changes with time. Our initial attempts floundered because we didn't include birth and death rates that *change* in time. Then, we corrected the problem by using an equation that included a complicated term that had K in it. Now, we are quickly confronted with the same conundrum. K *too* changes in time, or to put it in a less sciency way: the maximum amount of individuals that can live in an environment tends to go up and down. One year, the local woods might be able to support a max of 500 deer, but the next year some kind of blight might go through and kill most of the vegetation, dropping the maximum number to, say, 100 or so. This is reflected in the graph by a drop in the dotted horizontal line. The curve that represents the population then takes a dive in order to catch up with this change.

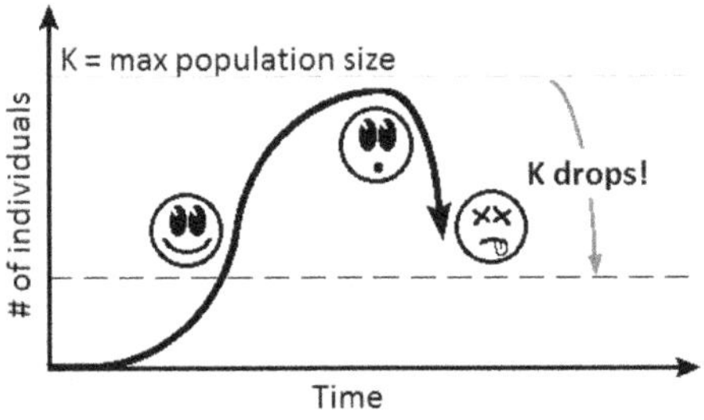

Common causes of such population crashes are severe unexpected weather, fires, sudden floods, volcanic eruptions and large scale human alterations of the environment, but I suppose they could also include meteors, alien invasions, the advent of reality TV, and the end of the Mayan calendar.[15] Deaths resulting from these kinds of things are often referred to as **density-independent**, because they are not caused by the density of the population.

Density-independent processes don't necessarily have to be extreme, despite the pseudo-ludicrous list I just gave you. Often they are an inherent characteristic of an environment, so much so that it's probably safer to think of the population limits (the K we talked about) as being fundamentally variable. After all, nothing stays the same forever. A slight change in water availability from one year to the next could spark a cascade of events that drops the maximum deer population from 500 to 450, or increase it from 500 to 550, say. It's subtle but present nonetheless.

This fundamental variability in the environment has led to a number of different evolutionary strategies in organisms. I'll point out a few of the more distinctive patterns, though I will hazard a warning that these come with a fair amount of hand-waving. Exceptions occur, after all.

Tiny animals tend to reproduce faster

Being small has its advantages, but being able to resist changes in the environment *isn't* one of them. Since environmental

---

[15] I live under a rock, but as I understand it: the Mayans used up all their paper while making their really long calendar. The shortage of paper was supposed to cause the world to end somehow. According to experts, this end-of-the-world publishing oversight was supposed to occur before *these* bio books were to be published. In that regard, it's unfortunate that the world didn't end; one publishing disaster would have helped to avoid another.

change is ironically one of the few constants in reality, many small organisms reproduce rapidly – if not explosively – as a countermeasure to unexpected mass death. This is usually most obvious in insects, where a single brood might lead to millions of newborn bugs. The vast majority of these die before making it to reproductive age. For those that survive, life is a ticking time bomb, making quick reproduction all the more imperative. In terms of the mathematics we talked about previously, small animals tend to have a very high r_{max} value. Organisms that follow this pattern might get wiped out easily, but since they reproduce so quickly, they can make a rapid comeback.

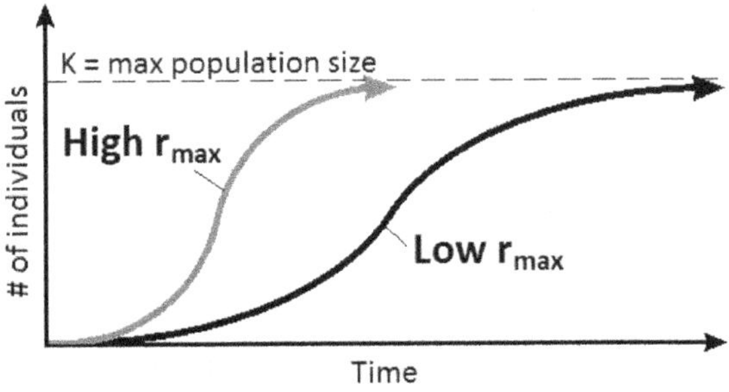

Large animals tend to reproduce slower

Larger animals take much longer time to grow than small animals, but they are much more resistant to environmental changes. Whereas a quick frost might kill legions of mosquitos, it will be little more than an uncomfortable chill for deer. As such, larger animals do not need to compensate for unexpected mass death by reproducing crazy-fast. They can take their time. In fact, since it sometimes takes a very long time to grow, parents often have to focus large amounts of time and energy rearing offspring. As such, it's usually more beneficial for them to have few children and focus their resources on them. The r_{max} for such organisms is very low.

Although organisms that follow the low r_{max} lifestyle might be more stable against adversity, there are disadvantages. Their low reproductive rate means that their populations do not recover very easily from disaster. Sometimes this fact becomes painfully clear after we use pesticides to wipe out pests, leading to a drastic decrease in the bigger predators.[16] The pests can make a rapid recovery, whereas the larger predators may take a long time to bolster their numbers.

These two reproductive patterns are sometimes described in terms of survivorship curves, which represent the probability of surviving vs. age, as shown here:

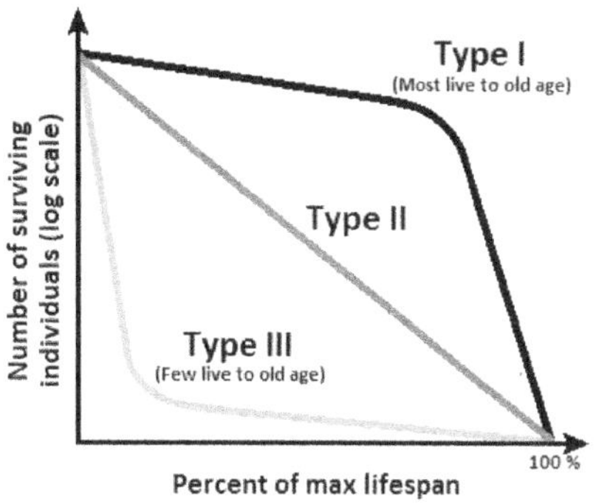

You'll notice in the image that I've drawn three curves, labeled Type I, II and III. The Type I curves represent organisms that generally reproduce slowly and live to old age. Many human populations – especially those of Western societies – are represented by this curve. Their lifespans are long, and most children survive into late adulthood as a result of improved living conditions, better nutrition, the relative lack of predators, and advances in medical technology. When the conditions are right, we linger to a ripe and grumpy old age – much to the dismay of everything else.

The Type III curve represents organisms that produce a lot of young, few surviving to old age. The example of the insects I mentioned above is an extreme version of Type III, but organisms that follow this pattern are not necessarily all tiny. In fact, most animals

---

16 ... since the pests are the food source for the predators. Kill the pests and the predators have nothing to eat!

follow type III curves. It's also very common among plants as well, which produce countless saplings that usually do not survive to adulthood. Type III curves teach us that being young is not always a good thing.

Type II represents those in the middle, which reproduce at an intermediate rate and whose death is not necessarily predictable based on age. Their lifespans are more favorable than Type III, but not quite as good as Type I. Examples include a number of birds, coral, hydra and biology book writers.

Changing Ecosystems

Just now, we talked about how environments can change, leading to alterations in the maximum number of individuals living there and to differences in reproductive strategy. Now, we look at the inverse situation in which the species living in an area change their environment. Specifically, we'll look at how plants and animals conquer newly exposed ground and how their actions transform the area. The result is a relatively predictable pattern of changes, called **succession**. There are two broad types of ecological succession: primary succession and secondary succession. Primary succession involves the deposition of new soil. Secondary succession involves conquering a region that already has soil laid down. It's probably easier to learn by example, so we'll start with the temperate land next to oceans. Let's head to the beach!

Primary Succession

Succession from Beaches

No body of water stays put forever. Oceans and lakes move over great periods of time, such as decades, centuries or longer, rivaling some of the slowest processes known: plate tectonics, mountain formation, tv commercials, etc. In the process of water recession, beach is exposed. If nature did nothing with the extra beaches that were left behind, the Earth would be littered with waterless-beaches. In reality, all this leftover ground becomes new real estate for plant growth. In fact, as soon as the water stops pelting that space, a number of species stand ready to pounce. On the other hand, the sand on beaches is not suitable for most land plants, so the first to move in tend to be small and sparse. In many cases, the plants that root there only live for part of the year, with each year bringing a new crop of annuals. With them come a few miscellaneous species of bugs, crustaceans and small rodents, (mice, rats, seasonal tourists, and so forth). It may not seem like much, but the importance of this austere stage in succession is that generations upon generations of little plants and animals slowly lay down a basic matting of **soil**. Soil is just a collection of plant and animal body parts that are gradually breaking down. It contains all kinds of debris and dead things from previous generations, like leaves, branches, bug limbs, bones and feces. All of these things are choke-full of nutrients that make it ideal for future plant growth. So, as the upper beaches acquire more and more soil, more robust plants are eventually able to take hold.

Further up on the beach, we encounter plants from an older stage in this process. This is where the beach *used to be* many years before, and here these plants usually live year-round. Their persistent nature serves to bind the sand and soil together, and they become the home for even more bugs, rodents and similar creatures. Behind this area of plant growth (and further up the beach) you'll find the first shrubberies and bushes. These build more bulk than the grasses and annuals closer to the water, and as a result they rapidly deposit more soil than any of the previous plant types.

Just beyond the bushes, you'll sometimes find a small patch of pine woods. Many pine trees and other related plants grow rather quickly, but are not as energetically efficient as the more advanced broadleaf plants. So, shortly after the pines begin rapidly growing, the hardwood broadleaf trees will sometimes move in. As these latter plants grow, they begin to shade and kill the pines. As such, just beyond the pine trees you'll see hardwood trees and a young forest. This last community of plants is the final stage of succession.

That is the characteristic pattern of succession: sparse annuals lead to sparse perennials, which give in to bushes and shrubs, which are taken over by small pines tree, which in turn are conquered by broadleaf forests. Walking from the water's edge toward the forest, you get a peek into the recent past.

Succession from Ponds

A similar pattern can be observed when a pond slowly recedes over decades.[17] Let's start with the center of the pond and move closer to shore. In the middle of the pond, it is too deep for any plants, except for algae. Cyanobacteria can grow here as well. These two organism groups grow and deposit sediment at the bottom as they die. This slowly makes the pond shallower.[18] As you get closer to the shore of the pond, you see a slow buildup of rugged vascular plants that are able to grow in shallow water. Next, nearest the shore, appear the broadleaf floating plants, such as lilies. These tend to shade and kill the deeper water vascular plants, restricting them to the inner parts of the pond. As generations of broadleaf floaters die, they contribute in turn to the further buildup of biological material at the bottom. Soon, this sediment makes the water shallow enough to allow the rigid cattails and other stalked plants to grow. This rapid

---

[17] Assuming more water isn't constantly added, which is sometimes the case.

[18] Those of you who are familiar with Archimedes Principle may point out that the water level should just rise if stuff is deposited at the bottom. You are quite correct, but increasing the water level will also spread the pond out, which acts to make it shallower. Also, and just as importantly, all that water is also *being used up* by the plants at the same time. This latter part is probably the more important process, leading to a gradual decrease in water depth and a decrease in the pond's diameter.

build-up of soil around the pond gives terrestrial plants a chance to take hold. When this happens, the water content typically drops low enough that the stalked plants are no longer able to grow. This gives way to grasses and other small land plants, which continue transforming the soil and decreasing water content. Further away from the pond, you see bushes and other shrubbery. Further still you see a small strip of pines and then hardwood trees farthest out, just like before.

Throughout this process, all the plant life is drawing water from the pond, and as we saw in Topic Text B, this water is used to make glucose. The glucose is then used – in part – to make cellulose. It is this cellulose that sinks to the bottom when plants die. So, in a sense, the plants turn the pond into dirt. Not all the cellulose sinks to the bottom of course. Some is eaten by animals, or carried away to make nests, or whatever. Either way, the water in the pond slowly recedes until the middle becomes nothing more than a marsh and then eventually just a meadow. When all of this is done, the plants that need lots of water die off and give way to plants that acquire water by pulling it from the soil. If you come back decades later, all that excess water will be gone and nothing will remain but a hardwood forest. If you come back even later, it will have become a parking deck, although I'm not entirely sure how that happens. In all my years of wisdom, I have begun to suspect humans for this last transformation.

Succession on Volcanic rocks and other newly exposed rock surfaces

Whenever natural forces leave rocky surfaces lying about, plants move in as quickly as possible. The first to take advantage of these new perches are lychens,[19] which can often take hold in tiny cracks. As they die, they deposit a buildup of dead material in the rocky depressions. Once the buildup is significant enough, mosses can start growing. As before, these also die and contribute to further buildup over many generations. Eventually, when the spot has been transformed enough, ferns and small vascular plants move in. Throughout this process, each generation chisels away at the rock's surface. This widens the cracks and breaks the rock down. The process can be rather slow depending on how hard the rock is, but you will eventually see larger and larger plants move in until the new rock surface is covered by soil and hardwood trees. At other times, some of the cracks of the new rock surface can be rather large, and within these cracks you see a rapid progression of succession, until a hardwood tree or two is growing from it.

Limits of Primary Succession

In all the above examples of primary succession, the main theme is the buildup of soil. This is the bottleneck in the process, because soil buildup can sometimes take decades to occur, depending on what other physical forces are at work. For instance, if wind and water are constantly beating down on a surface, that surface will experience slower soil buildup and greater runoff. As such, that area may take much longer to progress through the stages of succession. In the worst case scenario, it may not even progress at all.

Another factor that contributes to this process is the availability of water. In regions were water is lacking, such as in deserts, succession will not occur in this manner; rather, it usually stops with sparse and rugged vegetation, the kinds that are the most efficient at acquiring and holding water. Also relevant in the

---

[19] Well, actually the first to move in are bacteria. Bacteria are literally everywhere and *almost* as prevalent as coffee shops. To spot bacteria on newly cooled lava rock is not surprising.

discussion is the climate of the ecosystem we are considering. In fact, the above discussions are only generalized arch-types for succession. No two locations on the Earth are identical; the kinds of species that partake in the succession of one area may not be the same as those of another. Arctic areas, for example, rarely progress beyond the pine stage, as these plants happen to be particularly resistant to cold. In tropical areas, the number of plants and animals that participate may be so diverse and grow so explosively that it's difficult to argue that discrete stages exist at all.

Secondary Succession

In secondary succession, an area of earth is cleared of pre-existing and well-established vegetation, such as might happen if humans cut down a forest or if a fire rips through some woods. After the vegetation is cleared, you have a region that *already* has soil in it. In this case, what happens is usually a free-for-all for whichever plants are capable of moving in and growing the fastest. At first, you'll see grasses and bushes take over, as well as some pine trees. As the broadleaf trees begin to take hold, they will begin to shade the plants under them, which will then die off. After ten, twenty, or thirty years, the area will be covered in a hardwood forest again . . . ripe for the appearance of a parking lot or golf course.

As before, the above limiting factors of water and climate also apply, making the individual instances of secondary succession different from each other.

Climax Ecosystems

Regardless of how a hardwood forest gets to where it is, it does not usually progress through any other stage of succession. It is a **climax community,**[20] which means – broadly speaking – that the community does not change much from one generation to another. This should *not* be taken to mean that they don't change at all though! On the contrary, hardwood trees grow and die all the time. Fifty years from now, you will have a hard time recognizing the forest as it is today. Not only will the *placement* of trees change, but so will the *kinds* of hardwood trees, which do not necessarily follow any predictive pattern. A patch of hickory trees today might give way to a big elm in twenty years, which in turn might yield its space to an ash tree fifty years from now. Change is constant, but regardless of the smaller changes, a hardwood forest usually stays a hardwood forest. This is equally true of other kinds of climax communities. Cacti growing in a desert one year may be different from those of another year, but if you come back in fifty years, the region will probably still be a desert with cacti. Unless major environmental forces move in to alter the basic conditions, the climax ecosystem will likely stay there indefinitely.

Other Patterns of Succession

As we have seen, succession is marked by a gradual buildup of soil and slow progression to larger kinds of plants, where the final plant type is usually the hardwood[21]. Other patterns of succession can be seen as well. As succession gets underway, we often see a gradual increase in shade, a decrease in wind, and an increase in soil nutrient content. In addition, we typically see a rapid increase in the number

---

[20] The idea of a climax community is a very old one and has fallen into disfavor by many ecologists, mostly because of differences in details that exist between one community and another. Here, we are only interested in the prototype – the idealistic model, if you will – of succession. If nothing else, this discussion has historical relevance.

[21] . . . or whichever kind of vegetation is the dominant form for the climate in question.

species that take part in the transformation process. This process increases then slows as succession continues. The *kinds* of species changes as well. Then, once the climax ecosystem has arrived, the total number of plant and animal species usually stabilizes and doesn't change much.

As succession progresses, the amount of energy that is harvested by photosynthesis slowly increases as each stage's plants become more numerous and more efficient. When hardwoods take over, the efficiency in photosynthesis rapidly increases. Since photosynthesis results in the buildup of cellulose, we also see a drastic increase in the total biomass once hardwoods take over.

During succession, a gradual development of *vertical* ecosystems also occurs, especially as the climax ecosystem arrives. Much like the progression of human settlements, the first stages are near the ground, and then each successive stage gets taller and more complex. In the climax ecosystem such as a jungle or temperate forest, the vertical height of the ecosystem can be several hundred feet high. All of this altitude provides great opportunity for other plant and animal species to make their homes. This increases the complexity of microsystems within the forest. Complexity, in turn, tends to lead to greater stability of the species within the ecosystem. Let's have a look at that before closing up this unit.

Ecosystem Stability

We've just seen examples of how ecosystems tend to change. In the presence of sufficient moisture, bare areas slowly accrue more and more soil from generations of progressively larger plants. This leads almost certainly to a climax ecosystem, with many complex microsystems within it. We also joked around about how these turn into parking lots and golf courses, but transformations such as these can give us a segue into the factors that actually change ecosystems.

What factors make ecosystems resistant to change? What does it take to alter an ecosystem? What trends can we find?

Well, first and foremost, each ecosystem is unique with a unique collection of physical matter and a unique collection of species. Whatever may destabilize one ecosystem may not destabilize another. As with many areas of biology and ecology, there is an adequate amount of hand-waving. The topic can be very complex, so let's just skim the surface of ecosystem stability.

Early theories held that ecosystems that had lots of species in them were more resilient to change. The real trend seems to be the opposite. Communities with a *small* number of interacting species seem to be better able to withstand change, arriving at a new equilibrium shortly after the change is introduced. For the more species-numerous communities, repercussions resulting from an initial change can bounce around the ecosystem for generations before arriving at a new status-quo. On the other hand, ecosystems with increased physical complexity seem to lend to more species stability. For example, let's suppose you are attempting to avoid your professor because he has an assignment for you. In which environment would you be more successful at avoiding him: in a flat plain or a complex canyon with many hiding spots? Obviously, it is much easier to avoid a professor if you are in a canyon with oodles of cracks and crannies. If you are standing on a flat plain, the odds are good he will spot you, and then you're hosed. Likewise, animals that make their homes in complex physical environments seem more resilient to change than those who live in less complex areas. As always, the rule is a very soft and general one, so don't complain to me if the professor still finds you even after you hide in the canyon. If the strategy works, however, feel free to credit me for the idea.

Regardless of the complexity of the ecosystem, sometimes the loss of a single species can have profound effects on the remaining ecosystem. Such a key species is called a **keystone species**, and we can't always tell which species that is in advance. Think of it as the ecosystem's Achilles Heel. For example, let's suppose an insect species lives in an ecosystem – on a college campus maybe. Further, let's say that the insects bug us, so we develop a chemical to kill them. Unbeknownst to us, four other species are predators of the insect species: small mice, squirrels, frogs and bats. Further, these four species are preyed upon by other species, such as cats, dogs, freshmen and biology writers.[22] When we spray to kill the insects, all four small rodent species are nearly wiped out, which in turn drops the populations of cats, dogs, freshmen and biology writers. In this example, we would call the insect species a keystone species, because their removal caused disastrous change to the rest of the ecosystem.

Apart from killing off a keystone species or turning the place into a parking deck, how else might a stable ecosystem experience change? Well, generally speaking, anything that alters the average level of moisture, temperature, wind and other weather conditions has the potential to change an ecosystem, provided the change lasts long enough. In such an event, the species that take over are those who are best adapted to the new conditions. For example, suppose the Earth experiences global nuclear war. Many claim that the dust kicked up into the atmosphere by the nuclear ruckus would linger for years. If that's true, much less sunlight would reach the ground. This would drop the planet's total plant biomass. The drop in the biomass would result in a drop in herbivores, which would in turn lead to a drop in predators. When all is done, only small plants and animals would likely survive the event. After all, small plants have less structural overhead requiring energy, and small animals require less plants to survive. On the other hand, there is also the possible danger

[22] The latter two naturally capture small rodents in order to supplement their Raman noodle diet. It's like I always say: "A rodent a day keeps the doctor away, until you catch the plague."

that the oxygen content might fall dangerously low due to the rapid loss of plant life. If so, this could lead to the extinction of just about all forms of life except for the most devious and crafty organisms. . . there would be nothing left but bacteria and lawyers. It's a terrifying possibility, but it's also hard to say what would actually happen. Let's hope we never have a chance to test the theory out.

Summary

In this unit, we looked at many of the cause and effect factors of changing environments, as well as the factors that stabilize them.

- Environmental change is common, and when it occurs it often changes the maximum number of individuals within a species that it can sustain. This is denoted by a K that gets smaller. When this happens, any population whose size is higher than their K value will experience mass-death, until the population size drops below K. If K gets larger, the population will begin to increase in size until it approaches the new K value.
- Small animals are more sensitive to environmental change than large animals. To counteract the tendency of mass-death by environmental instability, small animals often adopt a high r_{max} reproductive strategy, which means they reproduce very quickly. They have lots of young and/or have very short childhoods. Some large animals (and most plants) also have a high r_{max}, so the "small species = fast reproduction" is not a perfect rule.
- Large animals have more resistance to environmental fluctuations. As such they can take their time to reproduce. They often have a low r_{max}, which just means they reproduce slowly. They have low numbers of offspring and/or long maturation periods. As with the "small species = fast reproduction rule," the "large species = slow reproduction" is also not written in stone. It is just a general tendency.
- The topic of reproductive strategies is often complemented by the presentation of survivorship curves. Type I organisms typically have low r_{max} values (reproduce slowly) and usually live to old age. Type II animals live to intermediate ages, and Type III tend to die very young. Type III species are often associated with having high r_{max} values.

- Whenever a physical region is laid bare, a series of ecological changes called succession will slowly transform the barren spot to an ecosystem that contains a dominant set of plant species that are appropriate for that climate. This final stage is often called the climax community. For deserts, it will usually be a community of cacti. For temperature regions, it will usually be deciduous hardwoods. For artic regions, it will often be a pine forest, and so forth.
- There are two types of succession: primary and secondary. Primary succession is slow and marked by the gradual buildup of soil. Secondary succession is rapid and characterized by a pre-existing layer of soil.
    - When lakes and ocean recede, the beach is first claimed by small sparse vegetation. These help build soil. Next to take over are more robust grasses that live year round (along with whatever insects and rodents make the area their home). Next to arrive are bushes and shrubberies. These give way to pine trees and finally to deciduous hardwood. These events imply an ample supply of moisture and appropriate temperature.
    - Ponds can often recede over the course of decades or centuries. From inside to outside, the species encountered are: bacteria and algae, underwater vascular plants, broadleaf floating plants, rigid cattails and other stalked plants, grasses, bushes, pines and the hardwoods. Each group lays down sediment, making the pond shallower and eventually making the conditions appropriate for the next group to follow. With enough time, the pond will disappear into a meadow and finally be taken over by the deciduous forest.

- When new rocks are exposed, lychens and bacteria will grow in the cracks. They deposit the initial soil. Soon mosses will begin to grow, followed by plants such as ferns. Eventually, vascular plants and hardwoods will take root. By this time, the entire surface of the rock will typically be covered in soil and plant growth.
- When a forest is cleared and laid barren, the vegetation will rapidly move back in, starting with small grasses and shrubs, followed by bushes and pines and then by hardwood trees. The progression is much faster than with primary succession.

- During the course of succession: shade increases, temperature fluctuation decreases, soil water moisture increases, wind drops and the number of species in the region gradually increases. Also, the net amount of energy captured by photosynthesis drastically increases, especially at the stage of the hardwood forest. Finally, the complexity of microenvironments slowly increases as well.
- Ecosystems that have a large number of interacting species seem less resilient to change. The initial disruption is sometimes slow to move through the ecosystem, via each species-to-species interaction. Sometimes the change will affect many generations before a new equilibrium is established.
- Ecosystems that have small numbers of interacting species seem to be a little more resilient to change. Sometimes the initial change can spark an initial and dramatic set of events, followed by a rapid return to equilibrium.
- The more physically complex an environment is the more stable any given species within the environment will tend to

be. This is especially true for prey species because they have more places from which to escape a predator.

- Sometimes, drastic change can occur in an environment if an important species is wiped out. Such a species is often called the keystone species, and its disappearance is often followed by the disappearance of many other species in the ecosystem.

Unit 3

Ecology III: Animal Behavior

In this section we cover:

1) Theoretical basis for social behavior
2) Social disinclination for: murder, cheating (theft), etc.
3) Altruism and self-sacrifice
4) Conditions causing social misbehaviors

Social Organization and Altruism

As we saw a few units earlier, living in crowded places has many disadvantages. Disease, parasitism, competition, roommates leaving moldy things in the sink and neighbors with lousy music can make your day rough. On the other hand, there are also many advantages to it as well, especially for those species that work together. Cooperation can turn a rag-tag band of organisms into an efficient machine. For instance, a cooperating group can form a strong defense against other groups, and they can better resist fluctuation in physical forces. Further, through the systematic use of scouts, more food can be found for all. In order for such cooperation to occur, however, the group must display some degree of **social behavior**. Consequently, we now turn our attention to this phenomenon. Studying this is *very* tricky though! It's fraught with hand-waving and mud-slinging, primarily because it's not always easy to nail down causes. Are social behaviors learned? Are they ingrained in our genetics? Evidence can be found to support either claim, but here our specific focus will be on genetic theories, because these fall under the purview of biology. Learned aspects of behavior are better covered in a course of psychology or sociology.

As we progress through this unit, please keep in mind that we are treading on theoretical grounds developed through *many* observations of animal societies, *especially* humans. What follows should not be read as scientifically proven or as an indication that cultural learning is unimportant in the final product of social behavior.

A genetic basis?

There are many social species out there in the biosphere, and it's very easy to argue that each one lives in an environment that is fundamentally different from the others. Still, there are clear behavioral patterns that seem to emerge from each population, regardless of their 'culture' and upbringing. The mere existence of similar social patterns across such culturally diverse groups lends a strong indication for the genetic basis of social behavior. Perhaps a better way to state this is as follows: in order for a species to be social, it seems that it must first have a *genetic tendency* to do so. Super, but what does *that* mean? Well, under standard conditions the social organism will tend to behave in a manner that will foster group stability, and it doesn't always need to be taught how to do it.

'*Under standard conditions*' is the critical part to understand. No matter how strong the patterns, *everyone* has their limit before they snap. I point this out explicitly, because one of the first counter-arguments to these theories is usually something like: "Yes, but how about when so-and-so killed all those people?" *Genetic tendency* just establishes favorable odds. In a society of millions of people, with camera crews ready to pounce and an internet that shoots out information like a machine gun, even .01% social deviation would make it seem as though our society is falling apart. On the other hand, if only .01% deviated, then it would mean that 99.99% conformed. In any other area of science, numbers like that would be highly indicative that a theory was not far from the mark. At any rate, we'll get more into social misbehavior a little later, when "genetic tendency" will hopefully make more sense.

At the heart of the evolution of behavior is the notion of the **cost/benefit analysis**. If an action costs more than the benefit, then organisms that have a genetic tendency to choose that action will be at a statistical *disadvantage* in survival and reproduction. If the action in question costs very little but has great benefit, then those with the genetic tendency to take such an action will be at a statistical *advantage*.

For example, let's suppose there exists a species of big purple balloon-like creatures. Throughout the course of millions of years, let's say that anytime a different species encountered these balloons, two outcomes were possible. If the animal species was nice to the balloons, the balloons would give them food. If the species was mean to the balloons, the balloons would pop and release a deadly toxin that killed the rude animal. After millions and millions of years, the animals that were genetically prone to automatically be mean to the balloons would tend to get killed off, while those with the genetic tendency to be nice to the balloons would increase in number. Eventually, the animal species would simply 'know' at birth to be nice to the big purple balloons, because very few of those who 'knew' to be mean survived. The nature of the balloons would be recorded in their DNA, much like the nature of gravity, water and so forth. Like our basic ability to smile, niceness to big purple balloons would not need to be taught to offspring.

In the above example, the evolved species would have a genetic tendency to be social toward the big purple balloons. In that example, being nice carried little cost and yielded great benefit. Being mean had little benefit and great cost. Now, let's take on a more realistic example. Instead of balloons, let's use other members of the same species. In other words, let's say you have a number of options when encountering another of your kind out in the wild. What does evolution have to say about millions of years of such interactions? Would there be a genetic tendency to choose one action over another?

I suppose the answer depends on the species, so let's pick on a few. Let's say you are a Great White Shark. You are the biggest, baddest dude in the ocean. You don't particularly need help to survive. You take as you please and go where you like. When you encounter another Great White Shark, you have very little to gain by befriending the other shark. In fact, the other shark is probably just going to get in the way. On the other hand, attacking the other shark isn't necessarily a good idea either, because you could be seriously hurt or killed in the process. The best course of action is to swim away and ignore the other shark. Cost/benefit analysis would lead to a relatively reclusive species, since those who chose this non-confrontational path tended to survive with the least cost and highest benefit.

What about the lanky, goofy and slow human species? You certainly are *not* the baddest dude in the jungle. You can't outrun the lions or the deer, and you're often stuck eating twigs and berries. So, what is the cost/benefit analysis of the outcomes when confronting another member of your species? Suppose you both had the genetic tendency to attack each other on sight. In that case, one or both of you could get hurt or killed, and neither of you would be any closer to catching a deer or defending against a lion. You would face an uncertain future with possible protein malnutrition. What if the genetic tendency was to ignore each other? In that case, you could at least avoid getting hurt, but you still wouldn't have much to gain. The last option seems to offer the greatest benefit. Those who are genetically inclined to cooperate stand to gain a far more bountiful future.

Ok, great, so 'cooperation' genes win the gamble and after millions of years you tend to naturally clump into groups. What other genetic tendencies would need to be in place to strengthen the social interactions? Let's look at a few.

1. *Do not murder your tribesmen*

While it *is* true that murdering another member of your social group (tribe, pack, whatever) *will* reduce your competition for food, mates and resources, it *also* leads to inherent instability in the group. After all, if you're genetically inclined to kill other people, then others might be just as inclined to kill you. You might have a better pick of mates, say, but you would have to spend a great deal of energy watching your back. The constant expenditure of energy watching each other closely would reduce the effectiveness of the group, leading to lowered mutual defense, decreased regulation of physical forces, less efficient scouts and so forth. In effect, murder is just a step backwards to where we were before: no help in catching deer or defending against lions, and nothing to eat but twigs and berries.

So, while the benefit of murder would be *less competition*, the cost would be great *social instability* for the entire group. Individuals whose genes made them inclined to commit murder were less likely to work effectively in groups and were much more prone to die alone. As such, under standard conditions, millions of years of cost/benefit analysis would lead to humans that rarely commit murder.

> Genetic Social Rule #1:
>
> Do not kill other tribesmen unless you absolutely have to, because it 1) leads to increased risk of reciprocal injury or death and 2) it undermines the effectiveness of the group working together. Ineffective groups threaten everyone's survival.

2. *Cheating (i.e. theft)*

In order for the group to succeed, each individual's work must yield a positive gain to the group. Consider a small factory of 25 people. Suppose each day of work nets five dollars per person. Altogether, the group makes 25 x \$5 = \$125 per day. Suppose you are cheating and consuming more than you make;[23] let's say each day you consume three dollars. The group's total earnings would drop to \$117 dollars per day.[24] Your cheating makes the group less effective. On the other hand, the group is *still* making a profit, and as long as they don't catch you, your actions will benefit *you*. This is a slippery slope though, because if everyone did it, then the group would be wildly inefficient. Clearly, there must be a genetic tendency not to cheat. . . one that is less absolute than the tendency not to commit murder.

Ok. Let's say your neighbor cheated and got caught. What should be done about it?

It's easy to argue that the group should just cast your neighbor out,[25] but the confrontation with him will cost you energy, and he might fight back! That means you risk injury and death in confronting him. Not only that, but just because he cheated this time, it doesn't mean he will not be helpful later. So, whether or not he is confronted (and punished) depends on the cost/benefit of the confrontation. If he stole a dime, it's not worth doing anything about it. The hassle of confrontation would cost you more than the benefit. If he stole a week's worth of food from your child, that's different. When all is said and done, the genetic tendency of social groups is to have occasional small-impact theft. What do I mean by "small-impact theft?" Well, if you stole a million dollars from a village, you would practically cripple the community. If you stole a million dollars from

---

[23] Tisk Tisk!

[24] 5\$ -> -3\$ is a net loss of 8\$.

[25] . . . oh what a wonderful day that would be!

New York City, the city will barely be affected. Getting away with theft is pointless if your actions destroy the group; you'd be back to eating twigs and berries and catching your own deer!

Anyway, our second rule is:

> Genetic Social Rule #2:
>
> Cheating and stealing from your tribesmen should be discouraged as much as possible. Once in a while, the cost of theft will be low and the benefits great – leading to occasional theft. In catching cheaters, react only if the confrontation has low cost and great benefit.

The punch-line from this rule is that theft cannot be stomped out entirely. It will always be there, waiting for the right opportunity. Nature is always looking for opportunity, and your best defense is not to be idealistic. It's like the old saying: *everyone* has their price.

3. *Mating Order and Social Structure*

Mating is vital to the continued propagation of genes. On the other hand, the mating rituals of countless species show very high costs. Once again the cost/benefit analysis comes into play, leading to a very tricky balancing act. For example, it does you very little good to spend all your energy shooting for the 'hotty' in the group if she's clearly out of your league. After all, she's *also* looking for the best genes in her potential mate. So, no matter how mad your video game skills may be, she's probably not going to think that your top score in *Space Boogers from Mars* is a reflection of good genes. Nor will she believe that your level 78 dark-elf time-wizard/knight-captain from the mystical land of Mugpooyoo is an indication of sperm prowess. So, while the benefit of acquiring the perfect 'hotty' may be great, the cost of a completely futile attempt will be greater. You may have to settle for the next-best mate... or risk having no mate at all. The same happens in other animal populations.[26] The net result is a tendency for animal species to practice very structured, tiered mating groups. The highest-desired individuals mate with each other, the next-highest will pair with each other, and so forth all the way down to the genetically weakest, with several members of my distant family below that.

Many species establish mating pairs or groups that rarely change from year to year. Why? Simple, it saves a lot of time and energy. Seriously: flowers, dinner and a bottle of wine can get expensive. It is often even more costly to attempt to steal a mate from an established mating pair or group. In the worst case scenario, you may end up in a fatal or disabling confrontation with the mate-defending rival. The best case scenario, in contrast, is usually the least likely; your rival will probably not back away without physical rebuff ... and yes, he already knows how awesome your time-wizard is. So, to put it bluntly, trying to take your neighbor's wife is rarely genetically efficient... under *standard* conditions, that is.

> Genetic Social Rule #3:
>
> In the vast majority of cases, leave your neighbor's mate alone. Unless the circumstances yield low cost and great benefit, you will likely not gain from the attempt. Those in the past who tried usually spent too much with little or no gain, resulting in less offspring.

[26] . . . except that *their* dark-elf time-wizards totally suck.

4. *Altruism*

The first three rules give us the genetic tendency not to do certain bad things to each other. Society is more than the sum of don't-do-bad-things though. We clearly do good things for each other as well. How can we explain this? We begin, in part, by a concept known as altruism.

Altruism is a term that refers to the social perception of 'selfless giving.' It often appears in the form of helping strangers, such as holding doors open, picking up someone's coat, and so forth. Many people consider altruism as strong evidence *against* evolutionary theory because of the belief that 'do-gooders' have nothing to gain from the action. Surprisingly, the reality appears to be the opposite. It takes a bit of mental contortion to understand the arguments though, so bear with me. Again, these are not proven theories, so don't get your knickers in a bind.

As before, we start from the genetic cost/benefit analysis. In the case of altruism what we often find is that the most costly acts of altruism are reserved for those we *perceive* to be closest to us. If we perceive that someone is close to us, we are more likely to loan them our car, our laptop, our money or whatever. The less familiar someone is, the more reserved we are. For instance, opening the door for a complete stranger carries practically no cost, as does picking up their coat. The benefit of these actions is that you may look more socially acceptable to a potential mate. Also, you may benefit from reciprocal altruism in the future. In effect, opening the door for someone is an advertisement of your willingness to be a "team player." On the other hand, you aren't quite as likely to commit high-cost altruism on a stranger. You wouldn't, for example, give them a key to your house in case they needed a place to rest. The benefit of this action is vague at best, and the costs could be grave.

Why are we more helpful to those closest to us? Well, in our distant past, we lived in small tribes and the genes that led to social altruism could easily have been shared by numerous members of each tribe. When an altruism gene caused one person to help another, the odds were good that the recipient of the 'do-good' action also carried a copy of the *same* altruism gene. Since that person had the same gene, he or she would also be more likely to help in return. This interaction increased the collective odds of survival for those with altruism genes. Over the course of millions of years, this amplification could cause altruism genes to become extremely common.

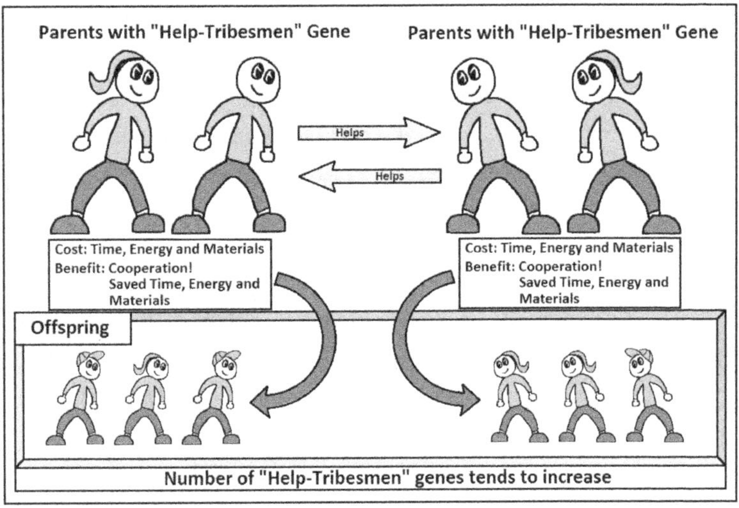

That said, this little trick of genes helping identical copies *can* be a gamble, because there is technically no absolute guarantee that another member of the population *definitely* has an identical altruism gene. So, those genes that focused altruism preferentially on close relatives tended to win the bet, because those who are genetically related are more likely to share the same copies of genes.

"Oh sure, Dr. Ickes, but why do we help our friends when we *clearly* know they are not genetically related? Answer that smarty pants!" Ok, simmer down. There is no need to attack the putative intellectual attributes of my trousers. We're getting a little deeper into theory. To properly understand this phenomenon, it's important to put yourself in the position of an altruism gene in a small low-technology tribe. You'd *like* to focus on close genetic relatives, but you have no eyes or ears; you have no lab tests. You have no choice but to rely on the perceptions of the individual you are in. Think of it this way: if you feel "close to" somebody, then it usually means 1) there is a history of successful social interaction with that person, and 2) since you live in a small tribe,[27] there is a *very* good chance the person has genes in common with you. After millions of years, the feeling of "being close to someone" becomes an excellent first-line approximation for genetic relatedness and/or social reliability. If nothing else, it labels that person as having good "social credit." Helping them is low cost/high benefit. In contrast, before the advent of science, *knowing* genetic relatedness was not always reliable or even possible[28] - unless the other person was *really* different, in which case you could definitively say they were not related. Feeling close to someone *and* having a strong *hunch* they are genetically related results in the lowest cost/benefit analysis.[29] The odds are very good

---

[27] Our genes don't know we live in communities of millions of people! Shhh! I won't say anything if you won't.

[28] Except for mother-child, few family relationships carry guarantees of actual genetic relatedness. Even siblings can have very little genes in common. The mother-child bond is often cited as one of the strongest familial bonds in human nature, and this can be understood in light of certainty in genetic relatedness. It is one of the few instances in which one person *knows* without a doubt that another is related: the mother carried and gave birth to the child. For the child, in contrast, the relationship is not technically certain. For example, they could have been stolen at birth. This is rare though, so usually the child reciprocates an equivalent bond. The father-child bond, on the other hand, is sometimes less strong, because there is always the possibility that the kid could be from some other male. All aside, family bonds still carry a much higher possibility of true genetic relatedness. The closer someone is to someone else on a family tree, the more likely the odds of altruism.

[29] There is, of course, the possibility of *not* being close to someone and yet *knowing* they are closely related. You might help such a person a little more than you would the guy on the street, but you still wouldn't extend your fullest altruism to them. This is one way in which the genetic gamble that "feeling close to someone" = "they are probably related" breaks down.

that helping these individuals will pay off very well, because for millions of years your social genes have benefited from such gambles. So, here is rule #4:

> Genetic Social Rule #4:
>
> Help those you feel close to, *especially* if you believe they are close family. The greater the degree of relatedness, the greater the odds they carry the same 'helping genes' as you. Even if they are not related, they've proven themselves to be reliable and will be likely to reciprocate help if you need it in the future.

5. *Self-Sacrifice*

The existence of self-sacrifice – or giving up one's life to save others – is another aspect of social behavior that many point to as a strong argument against evolutionary theory. This behavior, however, can be lumped in as an extreme example of altruism. In many cases of self-sacrifice in the animal world, what we sometimes see is that the oldest members of a family unit will typically be the ones who commit the act. For example, in prairie dogs, one member of the family (usually the eldest female) will stand watch for predators. If a predator is spotted that appears to be moving toward the group, the female will issue a loud call. This naturally will get the attention of the predator, sometimes leading to the death of the female. Here we can understand this otherwise perplexing predator warning by realizing that the genes that lead to this behavior are probably shared by the others in the family. One gene saves ten other copies of itself. Further, since the eldest female is beyond reproductive age, it maximizes the ability of the other altruism genes to reproduce by preserving the reproductive-age members.

Another very striking example of this can be found in any insect species in which all reproductive potential is focused on the queen of a hive. In these species, the brood creates thousands of sterile workers, all of which will readily die to protect the queen. All these offspring are undeniably the children of the queen. The genetic relatedness is beyond question. Since the drones cannot reproduce – and because the queen is carrying the same genes – they contribute to the survival of their genes by dying in defense of the queen.

> Genetic Social Rule: #5
>
> If increasing the numbers of the altruistic gene requires sacrifice of one copy, then the action – no matter how regrettable it *may* seem – is warranted. Copies that exist in bodies that are beyond reproductive age or in bodies that cannot reproduce are the best suited for this role.

6. *Us vs. Them*

When we began talking about altruism, we noted the idea of the perception of relatedness. In theory, altruism and social genes work by amplifying their numbers through repeated reciprocal assistance to each other. To do so, they must rely on their perception of closeness to other people. The greater the feeling of closeness, the greater the odds of committing acts of altruism. What about in the opposite extreme? What about a stranger that shares no cultural, religious, political, linguistic or physical features with you? In a small tribe, these things are strong indications of non-relatedness. Without genes in common, altruism genes would be spending a great deal of energy and materials helping a stranger with no guarantee of benefit. This expenditure of resources results in a disadvantage to reproduce, lowering the survival probability of such a generous gene. In other words, there is no inherent tendency for these genes to amplify, leading to the genetic tendency for xenophobia in biological societies.

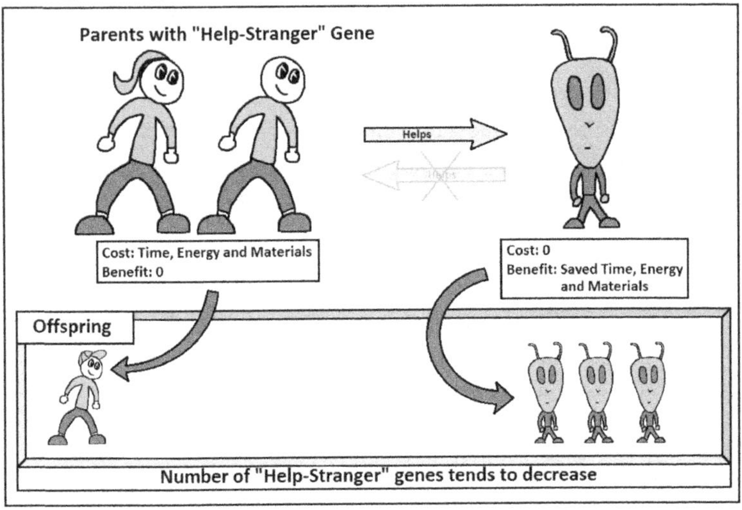

The first five social rules apply to actions toward fellow tribesmen: Do not commit murder *to your tribesmen*. Do not steal *from your tribesmen*, etc. When it comes to a complete stranger, these rules become less certain. The stranger may appear threatening to you. Alternately, they may show very little sign of interest in cooperating, and/or may represent clear competition for survival. Violence, especially in light of linguistic differences, is often regrettably unavoidable. In the best-case, and thankfully the most common scenario, you and the stranger will regard each other with suspicion and will avoid breaking the above rules in order prevent escalation of violence. In this event, small low cost acts of altruism usually occur as a token of willingness to negotiate. In effect, it's a way to test the waters to see if the other individual has compatible altruism and social genes.

Genetic Social Rule: #6

If someone is clearly not from your tribe, then adherence to rules #1 through #5 is not required ... except as a way to avoid harmful conflict. Proceed with caution, perhaps by extending low cost altruism as a test.

Distribution of Expression

All physical traits have distributions. Among brown-haired people, for instance, a range of colors can be found, from the nearly-blond to the nearly-black. If behaviors are rooted in genetic tendency, then they too will show a distribution of expression. There will be those who follow the genetic tendencies without fail and those who skim the edge. There will even be those who blatantly ignore the rules altogether. Most will adhere.

What might lead to such a distribution in behavior? Well, for starters, a mutation in any gene responsible for the basis of social behavior could result in abnormal behavior. Such individuals might be 'a little off' or might appear mentally ill.[30] Alternately, people living through horrific experiences may no longer be able to distinguish standard from nonstandard conditions. Or, they may no longer be able to identify with any 'tribe.' Thoughts such as "no one understands what I experienced," or "I no longer belong here," may alter their perception of closeness to others and could put them at risk for straying from the genetic rules that would otherwise influence them to help other people. At other times, humans may break their genetic tendencies because they feel they have no other choice. In

[30] Conversely, just because you might think someone is a 'little off' or mentally ill, that is not an indication that they have some kind of mutation. It is best to leave the diagnosis of another person's behavior to someone specialized in such a field.

fact, I'd wager that 'receiving orders' from someone who miscalculated rule #6 is probably one of the biggest causes for social misbehavior. There are probably lots of reasons in which social behavior can break down. Human cultures are very complex, leaving lots of room for conjecture.

Regardless of the reasons, any behavior that leads to instability of a social group is often – and rather broadly – referred to a **sociopathic** behavior. We will not go much into depth, because this is covered more thoroughly in a course like psychology, except to say that the symptoms can range in severity from being perpetually rude, to locking yourself in a grungy apartment and writing biology books, to something much more severe like deliberately injuring or killing others. Some symptoms are obvious while others are not. In fact, not everyone will always agree that a certain behavior is sociopathic, which makes this area of study particularly murky.

Closing Remarks

When I first studied the genetic theories behind social behavior, I was immediately struck by how well they parallel those propagated by major human cultures. In this latter respect, the historical view is that such rules (rules of morality, if you will) were given to humanity by external forces. Behavioral ecology tells us the opposite: our perception of morality is merely a reflection of the internal drives that direct our social behaviors. These internal drives exist because without them we wouldn't have survived the tribal days. We put to writing what we feel is fair and then go to great lengths to teach such rules to our children, without realizing that most of these teachings may be unnecessary. If the theories are correct, then telling little Timmy not to kill other humans is probably as pointless as instructing him not to leap into a crocodile pit. In that light, I've never come upon someone having a discussion where the line "Sorry, madam, but my son is still in his murderous stage" was ever uttered. In all likeliness, I never will.

On the other hand, terrible things admittedly *do* happen, and people sometimes do *really* bad things. When these things happen, behavioral ecology offers us several logical places with which to analyze what went wrong. Does the offender feel a lack of (or damaged) kinship with the person wronged? Does the offender have a mental illness? Has the offender experienced a traumatic event? Were there extenuating circumstances? All these considerations help to make sense of our human behaviors. In contrast, simply labeling someone as 'bad' does not help to solve the situation nor prevent it from reoccurring. Such branding only serves to make things worse.

If we are to move forward as a species, understanding who we are – and how *similar* we are – is critically important. Millions of years of evolution have made it clear that teamwork is the way to succeed, and recent genetic evidence suggests that any two humans picked at random are about 99.9% genetically identical. That means we are much more related than we think. Sadly, our genes 'think' we live in small simple tribes, where less is to be gained by helping those who look different. The reality is much more upbeat: billions of people live on the planet, and we have achieved technological wonders beyond imagination. If we can work together we accomplish even more. We can go much further than catching deer and avoiding lions; we can cure diseases, establish bases on the Moon and on Mars, find cleaner fuel, and maybe even eradicate telemarketers ... The future is at our feet, but to get to it we'll need to reach passed our genes.

Summary

In this unit, we explored the basic genetic theories regarding the evolution of social behavior.

- The benefit of social behavior is that cooperation may result in greater benefit than if individuals worked alone. This is relevant for any species in which more is to be gained by cooperating.
- Social behavior shows several general patterns across almost all social species. Since these tendencies develop despite differences in culture, it suggests a genetic basis for at least some social behaviors.
- In order for a species to be social, it seems that it must first have a *genetic tendency* to do so under standard conditions.
- Practically all social behaviors can be understood in the context of a cost/benefit analysis. If something has low cost/high benefit, then it pays to carry through with the action. Conversely, if something has high cost and low benefit, it is best to avoid the action. In theory, cost/benefit analysis shapes the frequency of genes that give organisms the tendency to choose one way or the other. If one action consistently has high cost and low benefit, then genes influencing organisms to take this action will slowly drop in frequency over generations, due to the inherent disadvantage of taking that high cost/low benefit action. If another action is consistently low cost and high benefit, then genes causing organisms to take this action will increase in frequency.
- Cooperation requires stable social interaction, which fosters the following rules:
    - Do not murder your fellow tribesmen.
    - Do not steal from them, unless it is a minor theft and you can get away with it.

    - Do not challenge your tribesmen for his/her mate, unless the cost/benefit analysis of the confrontation is very favorable.
- Altruism is the perceived act of selflessness toward another individual.
    - High cost altruism is rare between two individuals that have no prior acquaintance.
    - High cost altruism is common between two individuals who have long-standing and positive history of social interaction, especially if both individuals believe they are genetically related.
    - In theory, altruism is due to genes that cause individuals to tend to help other members in their tribe, because the other members tend to have the same altruism genes. The other members then reciprocate. This reciprocal cooperation results in low cost high benefit interactions that tend to amplify the number of altruism genes from one generation to the next. Altruism genes "try" to target close relatives as much as possible, because close relatives have the highest chance of having the same genes (and thus ensuring the amplification phenomenon).
    - Altruism genes must rely on the perception of the organism carrying them. True genetic relatedness is often difficult to tell for certain, so a feeling of close kinship is a close approximation. Throughout the bulk of human evolution, we lived in small tribes. As such, a feeling of close kinship was usually an excellent indication of "social credit," with high potential for reciprocal altruism and/or altruism gene amplification.

    - Self-sacrifice is generally considered an extreme example of altruism, where one gene copy sacrifices itself to preserve many others. This often occurs within the eldest individuals or those incapable of reproduction.
- Xenophobic social behavior may have developed because major differences between individuals were a sure sign of genetic non-relatedness and thus represented a very poor chance of gene altruism amplification. Strangers are often offered low cost altruism as a test to see if they are capable of reciprocating. If so, the cost/benefit analysis of helping such individuals will become more favorable, but in theory never as favorable as someone you believe is genetically related.
- When social rules break down, they often do so between strangers (i.e. the xenophobic social behavior just mentioned). Other possible causes may be mutations in social genes (difficult to prove though), traumatic events that alter the perception of kinship, nonstandard conditions in which someone feels threatened, situations in which someone has no choice (receiving orders during war) and so forth.

Unit 4

Biogeography I

In this section we cover:

1) Wind and Rain patterns
2) Food Webs and Chains
3) Nutrient Cycles and Soil

So far, our discussion of ecology has focused on populations. We talked about how populations grow, how they are affected by their environments, how their environments are affected by them, and how they interact socially. Now we are going to transition to talking about more of the non-living components of the biosphere. Specifically, we're going to have a cursory look into a field called **biogeography**, which basically studies how biology and geography interact with each other. In this unit, we'll have a look at the driving factors behind rain and wind. We'll also trace the Sun's energy through biological populations and examine how elements such as carbon, oxygen, nitrogen and phosphorus are cycled through nature. Finally, near the end of this unit, we'll take a closer look at soil. In the next unit, we'll start to examine the major kinds of ecosystems on Earth.

The Planet's Electric Bill

Our planet is the third one from the Sun and gets practically all of its energy from it. When light hits Earth, the planet heats up and some of this heat gets radiated back into space.[31] The energy that

---

[31] Apparently, we do a lousy job insulating the place.

doesn't get radiated is stored as potential energy of many forms, like chemical bonds, evaporated water and so forth. The Sun powers nearly everything around you. If it's something you like, thank the Sun. If it's something you don't like, then blame it. The Sun is paramount to almost all life on Earth, and if it went out, our society would rapidly fade into a forgotten history, silently cursing our failure to pay the galactic electric bill. For so long as that great big bulb in the sky is burning, we'll keep getting energy, and that energy will continue to push strange natural phenomena.

Major Patterns of Wind and Rain

The Sun does not shine equally on all parts of the Earth. See, despite common belief, the world is round.[32] That means part of it faces the Sun while other parts face other directions. This leads to light that beats down strongest on the equatorial region of the planet and weakest at the poles. As we mentioned before, light energy makes heat; heat makes water evaporate. So, at the equator you tend to have a great deal of water rising up into the air. As air rises it cools. Cool air cannot hold water as well as warm air, causing it to lose its water, which turns into rain. This is why the equator receives a lot of rain. Further, the now-dry air drifts north and south because of the air still rising up below it. As it moves north and south, the dry air finally begins to fall back to Earth at roughly the latitude of the Tropics of Cancer and Capricorn, causing these regions to be unusually dry. If we look at a globe, we consequently see many of the world's driest areas at the tropic lines.

---

[32] This is an audacious claim, I know, but bear with me.

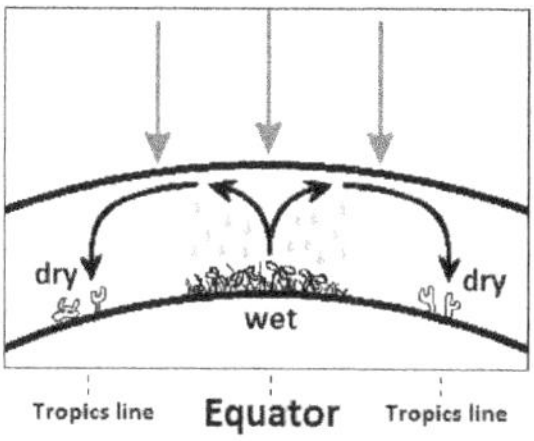

"Hogwash, Dr. Ickes! Not all of the tropic lines are dry! You're not tricking me with your Earth-is-round science mumbo-jumbo!" Very well, you are partially correct: not all of the regions of Tropics of Cancer and Capricorn are dry. The reason for this is due to wind patterns, which are complicated by numerous factors. Let's ease ourselves into this scuffle and get our feet wet with a discussion on ocean breezes as a first example. Land heats and cools much quicker than water.[33] During the day, the land tends to be much warmer than the ocean. This makes air above land warmer than air over water. It rises up and drags sea air in to fill the space left over. So, during the day, the wind comes in from the sea. At night, the land cools much quicker than water, so that the inverse tends to happen; breezes tend to go out to sea.

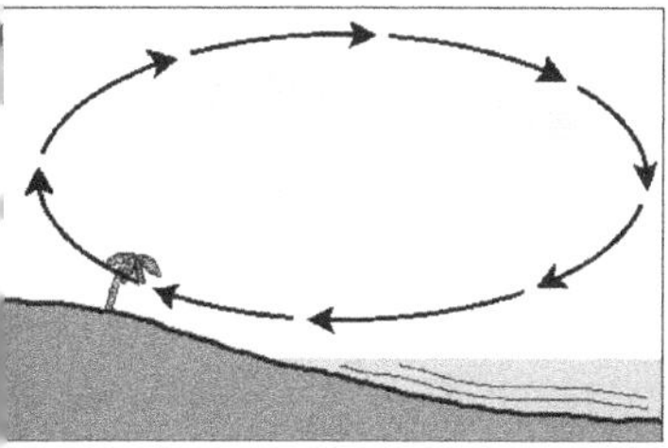

Sea breeze comes inland during the day.

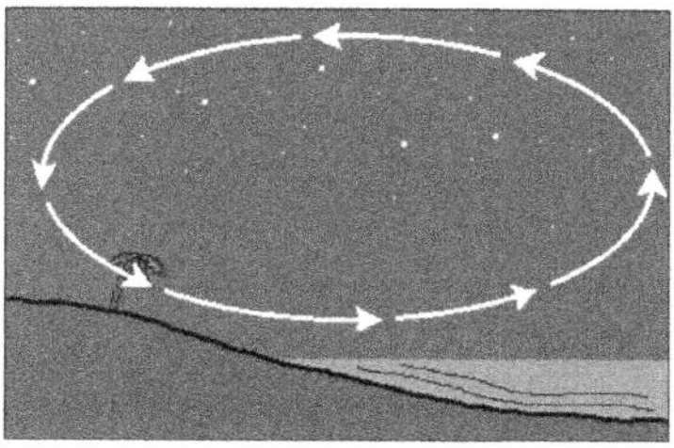

Breezes move out to sea at night.

Sea breezes tend to be a local effect and larger wind systems an mask them quite easily, if the conditions are right. So, let's zoom

[33] We saw this way back in our early days in biology. Water heats and cools slowly due o hydrogen bonds. This makes it a relatively good insulator against temperature uctuations.

out to a more global view. The Sun's light causes heat to accumulate. Heated air doesn't just rise; it also experiences an increase in pressure. High air pressure will cause net movement of air toward low pressure areas. This, however, is complicated by the rotation of the Earth. Wind at the equator is spinning at the same speed as the equator, but if it moves north or south it will encounter parts of the Earth that have slower rotational speeds. The difference in speeds causes wind to deflect in large circles hundreds of miles in diameter, a phenomenon known as the **Coriolis Effect**. So, if a high pressure area near the equator attempts to move toward a low pressure area elsewhere, it will begin to deflect and spin around the low pressure area, instead of reaching it. The further the pressure system moves toward the poles, the greater the spinning. This is the premise behind hurricanes.

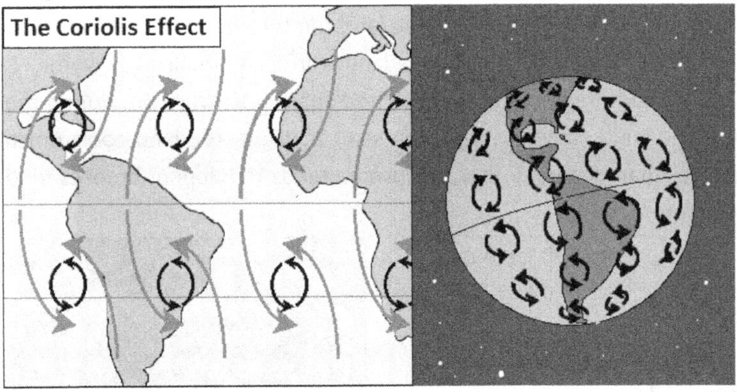
The Coriolis Effect

Another surprising culprit in the production of wind is the Moon. The Moon exerts a gravitational force. It may seem imperceptible to us, but to liquids and gasses the effect is significant enough to cause motion. Air and water move toward the part of the Earth that is nearest the Moon, leading to water currents,[34] tides and wind.

[34] This is not the only cause for water currents. In fact, many of the arguments made about wind also apply to water currents.

Finally, we must consider the placement of mountains. These rocky, physical barriers tend to influence the course of air currents by creating wind channels along the surface of the Earth. In addition, any time wind travels *over* a mountain range, it must rise and cool. The cool air dumps its moisture on the near side of the mountain. When it comes down over the top of the mountain, it is dry. As a result, the near side of the mountain is often wet and covered in vegetation, while the far side of the mountain is typically much drier than expected and sometimes covered by desert-like terrain.

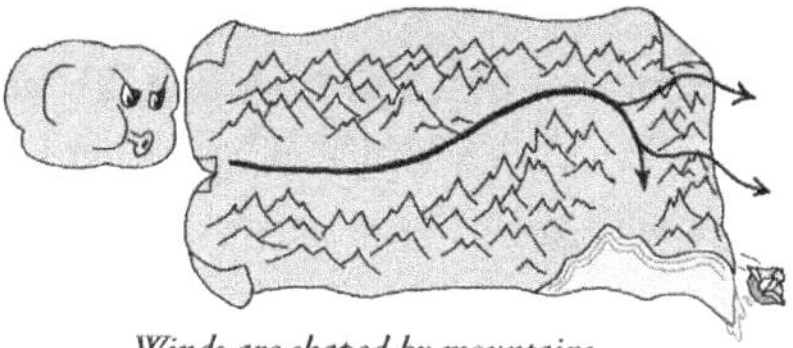

Winds are shaped by mountains. . .

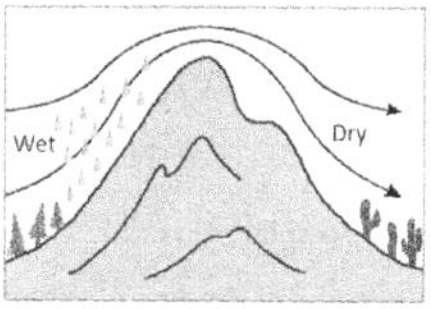

and drop their rain when climbing them.

We saw several examples of how rain is caused by rising air. When air cools, its ability to hold water lowers. The consequences to this can be observed on many hot and humid summer days when the late afternoon approaches and the temperature begins to drop. As the air cools, a quick rain can develop releasing much of the day's evaporated water back to the ground. Sometimes when a pocket of cold air is pushed into an area of warmer air, the warm air is rapidly pushed upward, resulting in strong rains. This is called a **cold front.**[35] In contrast, when warm air pockets encroach upon areas of cold air, the warm air gradually pushes upward in the process, resulting in prolonged slow rains. This is a **warm front.**

---

[35] Cold fronts are typically characterized by thunderstorms. The lightning that results is due to the separation of charge that develops between the sky and ground. It's more common in cold fronts because of the rapid manner in which the air rises, causing static electricity to build between all the dust, water particles and other molecules caught in the path. The details are outside the scope of this book.

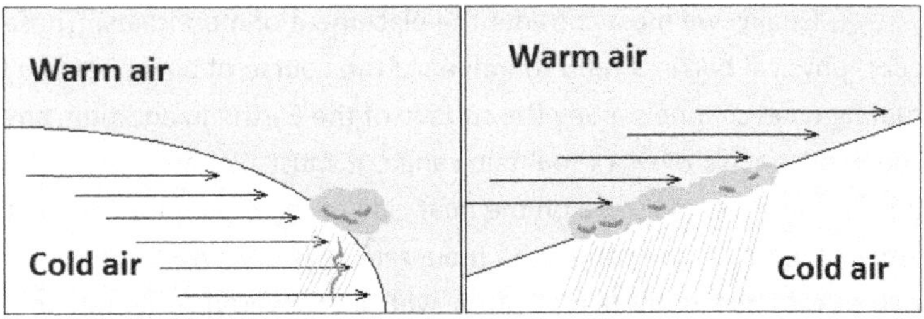

Cold Front **Warm Front**

The point of all this weather hocus-pocus is that wind gets its energy from the Sun and travels all around the globe in predictable but uneven patterns due to 1) the placement of water and land, 2) the Moon and 3) the rotation of the Earth. These winds drag moisture which gets released any time warm air cools. The rain that results causes deviation from the otherwise simple "dry Tropics of Cancer and Capricorn" principle I talked about previously. When all is said and done, the complexity of the wind currents leads to some regions receiving predictably high levels of rainfall and others very little. Since water is a critical component in life, the difference in moisture levels leads to striking contrasts between one ecosystem and another. In the next unit, we'll be looking a little deeper into that particular topic, but for now let's shift back to our discussion of the Sun's energy, focusing on the part trapped by photosynthesis.

Energy and Food Chains

Plants require light in order to do photosynthesis. On the other hand, light makes heat, so the areas of the Earth that receive the most light are often the warmest places as well. As such, photosynthesis and warmth usually go hand-in-hand, assuming enough moisture is present.

Like moisture, warmth tends to get moved around by wind. Despite this tendency, the dominant effect on temperature is latitude. At the poles, the sunlight barely makes a glancing blow across the planet. Near the equator, the sunlight beats down with all its fury.[36] So, the closer to the poles you go, the colder it gets. The closer to the equator you go,[37] the hotter it gets. Since photosynthesis follows light and warmth, you usually see a steady increase in photosynthesis as you move toward the equator.

Regardless of where a plant happens to be growing, it captures energy from the Sun and converts it to chemical form. Photosynthesis binds water and carbon dioxide together to form sugar molecules. Clumping all the Earth's photosynthesis together, the total stored energy is called the **gross primary productivity** of the biosphere. Part of this energy is immediately used up by plants during their own respiration.[38] The part left over is the **net primary productivity**, which is basically just a fancy way of saying "spare sugar." Plants take the excess sugar and make bigger molecules, like cellulose, oils and so forth. Animals eat the plants and make other molecules out of them as well. All this stuff eventually gets put into even larger things, such as trees, animals, cardboard, door-to-door salesmen, infomercials and so forth. It is all due to the net primary productivity – the excess energy that plants capture from the Sun.

Where does most of the net primary productivity occur? Not surprisingly, most of it comes from the rainforests, where the plant biomass is incredibly dense. Another large chunk comes from the oceans, specifically in the first meter or so of water. Why only the first meter? Well, water absorbs light quite readily, so the intensity of light drops considerably after the first meter or so. The deeper you go, the darker it tends to get and the less efficient photosynthesis is. Most of the photosynthesis in the oceans is done by cyanobacteria, which are

---

[36] Not that I blame it. It must be lonely and frustrating to be a star. Everyone is always watching you, but no one wants to get close enough to really know you. Poor star.

[37] . . .or more precisely: the closer you go to the place where the Sun is shining directly overhead . . .

[38] After all, do not forget that plants breathe too! They don't do photosynthesis to be nice.

present in high concentrations near the surface. The oceans cover the majority of the Earth. Still, despite such a massive surface area and huge numbers of cyanobacteria, surprisingly less photosynthesis occurs in the oceans than you would think. The reason is that many nutrients, such as phosphorus, are often the limiting factors in biotic growth. Occasionally human activity results in huge amounts of nutrients being dumped into the water. When this happens, you sometimes get temporary overgrowths of cyanobacteria, algae, dinoflagellates and other ocean critters. While this may sound like a good idea, these critters often release poisons into the water that kill fish.

Anyway, after rain forests and oceans, the next biggest "photosynthesizers" (in order of most to least) are the expansive broadleaf forests, boreal forests (taiga), savannas and then grasslands. Finally, places like deserts and the tundra bring up the rear. All these places are described in the next unit.

Regardless of where the primary productivity occurs, the products of photosynthesis serve as the base of energy from which all other organisms draw. This energy percolates through ecosystems in the form of energetic chemicals from plants to herbivores to carnivores in what is commonly referred to as a **food chain**. Each ecosystem has a number of food chains that crisscross their way from plants to carnivores in a **food web**. All along each chain are the decomposers. Decomposers are the fungi, bacteria, worms, bill collectors and other bottom-feeders that consume dead biological material. These organisms break down organic molecules into smaller ones, freeing a lot of CO_2, NH_3 and other simple molecules in the process. We'll return to these molecules shortly.[39]

At each link in the chain (a link being an individual), large amounts of energy is lost as heat. In fact, so much heat is lost along a

[39] That sounded ominous. I mean we'll be discussing these molecules soon.

food chain, that fewer and fewer individuals can ultimately be supported as you go higher in the food web - er, well, as a first approximation that is. So, whereas a geographic area might support 10,000 plants, say, these plants might only support 500 herbivores, which in turn might only support 10 carnivores. This scale-down effect is a common phenomenon, so much so that energy flow in ecosystems is often depicted using pyramid models that are divided into levels called **trophic levels**. The bottom of the pyramid is comprised of the first trophic level, occupied by the energy within **primary producers** (autotrophs). The second level, smaller than the first, is comprised of the energy within **primary consumers** (the herbivores). The third level, smaller still, is comprised of the energy within **secondary consumers** (the first level of carnivores). Many ecosystems have a fourth and fifth level, representing the carnivores that feed off other carnivores. Each of these levels becomes incredibly tiny, so much so that they aren't usually shown in the pyramid. The decomposers are also not shown because they feed off all levels.

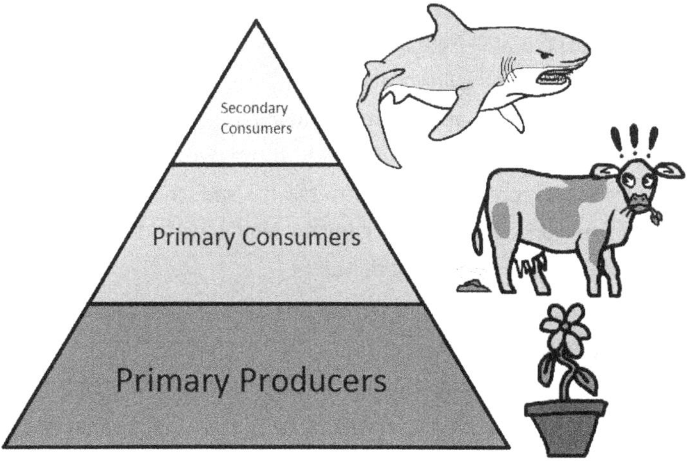

Let's clarify a bit. It is important to note that food web pyramids such as this do not necessarily represent the number of individuals – despite my example of 10,000 plants, 500 herbivores and so forth. Each level represents the total *energy* of those kinds of organisms. Why isn't the number of organisms a good measurement for the total energy? Well, consider an ecosystem comprised of one massive tree fed upon by forty thousand insects. In that case, there would be 1 plant, and 40,000 herbivores. That doesn't make a pyramid! If we take into account the total energy of the plant and the total energy of the bugs, the pyramid is restored.

Whereas these kinds of pyramids are not necessarily good representations of the total numbers of individuals within a given trophic level, they *can* be used to represent the total amount of biomass. If we return to our example of the big tree plagued by the swarm of insects, it should be readily clear that the tree could have a mass of several thousand kilograms, whereas the insects might have a combined mass of only a few hundred. So, a diagram representing the total biomass at each trophic level would be pyramid-shaped.

It's also important to note that many organisms can occupy multiple trophic levels. Consider humans. We eat plants, herbivores and many carnivores. Actually, an evening in front of the television will provide plenty of evidence that humans eat just about anything you put in front of them, including their last chance at dignity. As such, we occupy multiple trophic levels in the pyramid. In return, not much eats us, except for decomposers. It's sad; the only things that regularly feed on us are the organisms that have a taste for rotten stuff. Nature has us pegged all right.

Nutrient Recycling

We've seen how temperature and moisture cycles around the Earth, and we've also seen how energy trickles through an ecosystem. Now, we turn our attention to how basic molecules and atoms cycle around as well. In terms of matter, the Earth is a closed system. Until the second half of the 20th century, not much of anything ever left the

Earth and not much entered either. The consequence of this is that life must recycle the materials at its disposal. If it didn't, it would run out. All the oxygen, carbon, nitrogen, phosphorus and other basics materials must be put back after use.

The Water Cycle

Water rolls downhill whenever possible. When rain falls, it quickly makes its way into streams, lakes, rivers and finally into the oceans. At any point in this process, the Sun heats the water causing it to gradually evaporate. A small amount evaporates before it reaches the ocean. A larger fraction evaporates once it gets to the ocean.

Not all the water runs downhill. Some of it drips down through the soil and rocks. Plants pick up a lot of it as well. Once it gets into a plant, it is slowly drawn into the trunks and branches, and then eventually into the leaves. When it gets to the leaves, it either gets put into glucose via photosynthesis or it gets released as "plant sweat" through a process called **transpiration**. This process is studied more in detail in chapters or courses on botany. The highly anticipated blockbuster sequels to this book cover it in fascinating and heart-pounding detail...

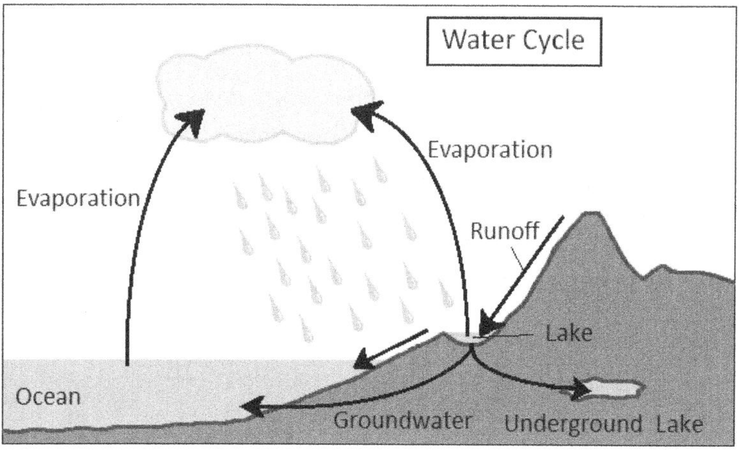

Anyway, the water that is not picked up by plants either dribbles into streams and oceans via underground routes, or gets stuck in deep underground lakes. You might think that all the water would eventually get trapped underground, but the deep earth can only hold so much moisture before it gets saturated. As such, we're in no danger of losing our water to the underground. Sorry Mole People, but you can't have all our H_2O!

The Carbon Cycle

The carbon cycle refers to the movement of carbon from the atmosphere to animal products and then back into the atmosphere. Based on CO_2's great stability and physical characteristics, as well as from the observation on other planets, we know that primitive Earth had a great deal of CO_2 in the air. Since the advent of photosynthesis, a lot of this carbon has been pulled from the sky by plants, specifically by the Calvin Cycle.[40] When plants die, they collapse and soon become buried by other plants that grow or fall on top. Over time, the plant material builds up. Some plants are eaten by animals, and the carbohydrates then get converted into various fats. When the animals die, they too tend to result in a build-up of dead material. Of course, not all the carbon gets stuck in dead bodies; a lot of it is returned to CO_2 by respiration.

Either way, imagine a world in which dead material is never broken down. The carbon in the atmosphere would slowly get captured and stuck in layer after layer of dead bodies. The CO_2 in the air would gradually drop until there wasn't enough for plants. Then they wouldn't be able to make their cell walls and would start to die. Then the animals would begin to die. Soon, you'd have a planet full of

---

[40] Yikes! Old material!

dead bodies.[41] Clearly, while some carbon *does* get stuck in the ground, we need to recycle as much as possible or life will stop.

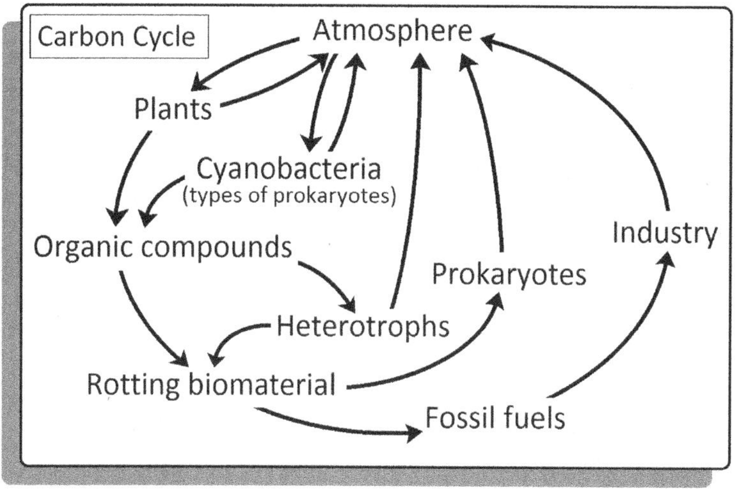

How do dead bodies get broken down? It's a complex topic, but the primary movers and shakers of this process are bacteria.[42] In fact – now that we're on the topic of bacteria – bacteria occupy two important points of the carbon cycle. First the cyanobacteria work to capture the CO_2 via photosynthesis. Along with plants, they fix a great deal of the CO_2 that is pulled from the sky. Second, a great deal of the remaining bacteria work to break down the dead, consuming the material and oxidizing the carbon back into CO_2, which then returns to the atmosphere. While not all the prokaryotes are involved, you can think of these simple organisms as the waste managers of the world. Without them, the process of life wouldn't have gotten very far.

You'll notice from the carbon cycle image above that humans play an important role through industry. We pull oil and other hydrocarbons from the ground (all of which are dead organic material

[41] Based on the current social trends, some argue that we've already reached that mark. I'm waiting for "America's Top Dead-guy" to air. Finally a show with a real pulse!
[42] And, yes, worms and other icky things are involved too!

from eons past) and burn it. In doing so, this releases the buried carbon back into the air. You might think that this is a good thing. After all, we wouldn't want it all to get stuck in the ground, right? Well, our role in the process is actually more destabilizing than helpful. When bacteria do it, they release CO_2 very carefully. When we do it, we release all manner of carbon compounds and soot into the air, rapidly changing the quality of the atmosphere. It is believed that these compounds have gradually begun to change the base temperature of the Earth (i.e. **global warming**), leading to more severe weather patterns and the melting of the polar ice caps.[43] Also, if the countless species on the planet do not have proper time to adapt to this rapid change in atmosphere, it may result in mass extinction.

Lastly, it's important to realize that we burn very *old* dead organic materials, not the new dead stuff. Despite our industrial habits, without bacteria we'd still find our world piling up with the dead.

The Nitrogen Cycle

Like carbon, a great deal of the nitrogen on primitive planets is usually trapped in their atmospheres, this time in the form of N_2. The reason is that N_2 is very stubbornly stable, and as such, not many of Earth's organisms are capable of pulling it from the air and recombining it into organic molecules. Bacteria are among the very few, and without them the other organisms would have a very hard time getting nitrogen. If you'll recall from earlier studies of biology,

[43] It's important to understand that global warming will lead to severe weather patterns, not necessarily just warmer average temperatures. Increases in storms and blizzards, ironically, are a symptom of this process.

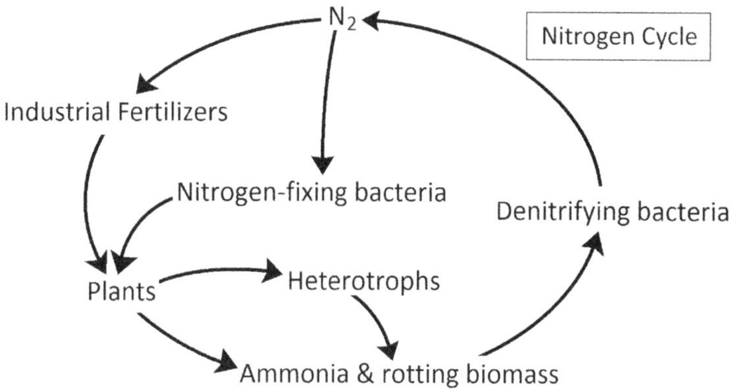

nitrogen is a pivotal component of amino acids and nucleic acids. If organisms do not have access to nitrogen, they cannot create their proteins or replicate/repair their DNA. Just like the carbon cycle, things would come to a screeching halt if bacteria weren't around to yank nitrogen from the air. This important ability is called **nitrogen fixing** (made possible by the enzyme **nitrogenase**), and those that can do it are called **nitrogen fixers**. On the other end of the spectrum are the **denitrifying bacteria**. These are the critters that return nitrogen back to the air by converting it back into N_2.

Oxygen Cycling

If you remember from previous studies, oxygen is happiest when it hogs the electrons from other atoms. As such, molecules like CO_2 and H_2O (water) are very stable because the oxygen atoms in them are immensely happy.[44] Conversely, O_2 is a *miserable* molecule, because neither of the oxygen atoms gets to hog electrons from the

[44] It may seem silly to say this, but atoms are not technically capable of happiness. I am merely describing it in a way that will hopefully be memorable. The physical and chemical reactions of reality take place because of differences in energy and chaos; emotional desires have no role, apart from those caused by humans and other misinformed organisms.

other. Left to themselves, all of these molecules would eventually react with the surrounding atoms to form more CO_2 and water. Our world would have no O_2 left to breathe! In addition to the automatic reactivity of oxygen, most organisms use it in respiration, returning it to CO_2 and H_2O.

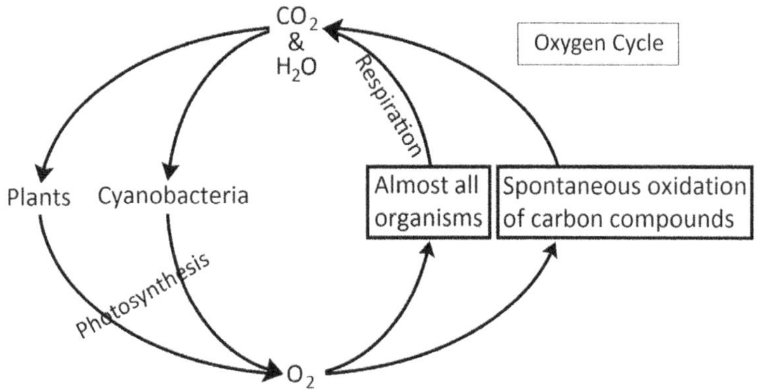

Prokaryotes (mostly cyanobacteria again) and plants play a major role in the cycling of O_2 by "fixing" it during the photosynthetic process. In fact, it is widely believed that cyanobacteria-like organisms were responsible for Earth's transition to an O_2-high atmosphere (see Topic Text D).

The Phosphorus Cycle

The primary reservoir for phosphorus is rocks. Billions of years of erosion have slowly etched it out from mountains and other rocky areas. Each time it rains a small amount of phosphorus trickles down into the streams and oceans. Once in the water, it is taken up by plants and animals where it gradually gets cycled through the marine and land ecosystems. Some of this phosphorus ultimately sinks to the bottom of the water or settles into the soil on land; it is then compacted by layers that form on top of it. Through geological time, these layers eventually get moved around by the many forces that act upon the Earth's land masses. Whenever new mountains are pushed up from below, the phosphorus that was compacted gets exposed again.

Phosphorus is often one of the many limiting nutrients in marine environments. It's one of the nutrients I mentioned before that sometimes get released when humans bumble around and drop stuff in the water. When this happens, large algal and dinoflagellate blooms develop, using up the oxygen and releasing toxins. This kills lots of fish and other marine life. Further, the drop in oxygen changes the chemical reactions that occur in the muck at the bottom of the water, which promotes the release of toxic gasses. Now that I think of it, this gives me a strong clue as to why my roommate's room smells the way it does, but that's a topic for a different day.

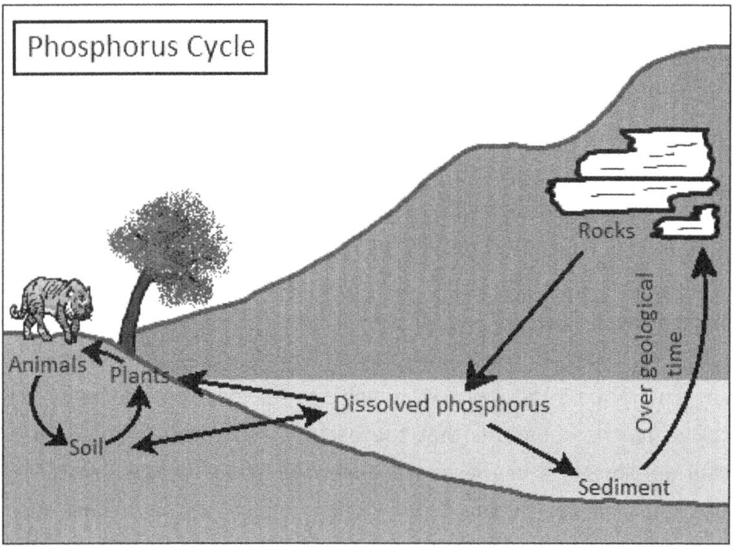

Novel Chemicals

In the modern world (far away from my apartment), new chemicals are being created at an unprecedented pace. They are created for so many things, such as novel industrials solvents, pharmaceutical chemicals, precursors to new plastics, pesticides and countless other things. Sadly, the rate at which these things are being created is outrunning our ability to safely test their effects on the environment. Sometimes these chemicals directly harm the ecosystem. Sometimes, the chemicals are harmless, but the ecosystem changes them into more harmful forms. I guess that's the price of technology. We might wipe out the planet in our attempt to find the perfect anti-pimple cream, but darn-it at least we'll look good doing it!

New chemicals can damage an ecosystem at any level. It can kill microscopic organisms, plants, insects, or even larger animals. The ways in which they do damage is a huge topic, and we could fill entire libraries with the details. A few simple facts bubble to the surface though. For instance, the damage that a novel chemical can do to its environment is not necessarily related to the amount released. One chemical might be many times more harmful than another, so that even tiny amounts will have devastating effects. In addition, we need to consider the phenomenon of **biological magnification**. In this process, a chemical that is released in trace quantities can sometimes be found in concentrations many *hundreds or thousands* of times higher inside some animals. How does this happen? Well, at the smallest level, bacteria and algae will take up a few molecules here and there. Many of these bacteria will be eaten by larger organisms, like protists or bugs, say. By doing so, the protists and bugs will acquire all the chemicals that the bacteria picked up, resulting in a small amplification in the concentration of the chemical. As larger organisms eat the smaller critters, the chemical gets amplified even further. By the time large animals enter into the food chain, the chemical has been greatly amplified. The higher the concentration within an organism, the greater is the chance that it will have an effect on its health. This is concerning to us, because the chemical can build inside the food that we eat – such as in fish, say. When we eat the fish, we get all the chemical magnification that happens to be there. Here is an illustration of the phenomenon.

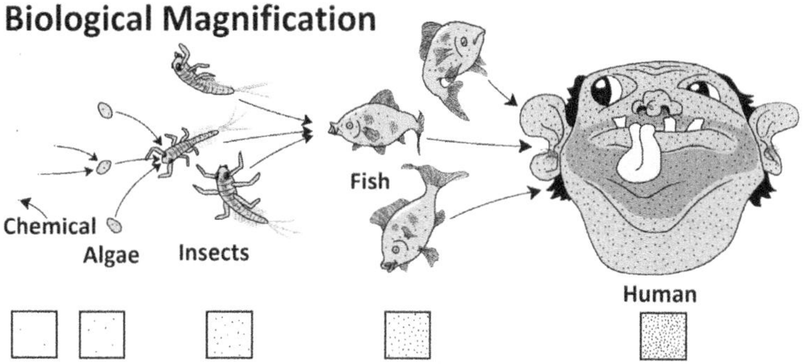

A classic example of biological amplification is DDT. DDT and its relative chemicals were used as insecticides in the agricultural industry, as well as controls for the spread of insect-borne diseases like malaria. Unfortunately, the chemical gathered in local streams and lakes, where it built up in concentration within fish. These were eaten by a number of birds and larger predators. DDT is believed to have resulted in the near-extinction of the American bald eagle, to name a few of the species affected. It is now banned for almost all but a few small-scale uses.

New chemicals can also alter the non-biological components of our environment as well. For example, fluorocarbons are used in a variety of processes, from cleaning and degreasing, to aerosol propellants, to refrigeration. Their use has dropped considerably, however, after it was discovered that they readily react with components of the ozone layer. The ozone layer is very important in the process of life because it blocks many of the harmful rays of sunlight. Continued use could result in a weakened ozone layer and subject practically all organisms to much higher potency of solar radiation. I suppose it is ok, I guess, because we'll soon have a cream for that too; all we have to do is whip up a few new chemicals.

The Soil

Now that we've looked at the wind, water and nutrient cycling, let's have a look at the basic characteristics of soil. We spoke briefly about soil when we talked about biological succession on beaches and newly exposed rocky terrain. Soil is critically important in the processes of life. First and foremost, it acts as a substrate to which plants anchor themselves. Without it, plants would just blow away, roll around and be subject to a great amount of physical damage.[45] Second, the soil acts as an important reservoir for nutrients and water. It keeps these things from quickly flowing away into the oceans.

Not all soils are equal. Soils have biological components and non-biological components. As discussed in ecological succession, the biological components of soil are the dead pieces of previous organisms. The more of this kind of material there is, the more the soil will tend to be able to support plant life, which accounts for the slow progression of succession. Pioneer plants will take root in sand and then gradually lay down dead material, which gradually increases the sand's life-supporting properties. More plants move in and so forth.

The non-biological components are typically comprised of variably-sized particles made from silicon- or aluminum-based compounds. Soils that have really small particles are typically referred to as **clay**. Mid-sized particles make up **silt**, and large particles comprise sand. The size and composition of the inorganic particles also affect its ability to hold water. For instance, in clay the tiny particles bind water and nutrients very tightly, making it too stingy for many plants to survive. In addition, these particles clump together into large globs, and new water flows off of them too quickly for plants to grab. When water rapidly flows away from soil, it tends to

[45] In case you are wondering, tumble-weeds are already dead.

drag precious nutrients with it, called **nutrient leeching**. If particle size gets too big, such as in sands, the water percolates through it rapidly, resulting in the same problem. Mid-sized particles seem more ideal for holding water and nutrients, although this alone is a gross over-simplification of the various qualities of soils. We aren't even scratching the surface and could easily fill up an entire course with dirt. Trust me; it's a very messy subject to dig through, but don't let it soil your opinion of the topic... Ok, I'll stop before I run aground.

Despite soil qualities, the process of nutrient leeching is also affected by the acidity of water entering it. For example, when acid rain falls on a soil, the H^+ ions will displace other cations, such as Ca^{2+}, Mg^{2+}, Fe^{2+}, K^+, Mn^{2+} and so forth from the soil's inorganic particles. Once dislodged, these compounds flow away and are no longer available for future generations of plants. In addition, the H+ ions can dislodge toxic ions such as Al^{3+}, which can kill plants and animals. Generally speaking, small amounts of acid rain are natural, but human activities have increased its severity by releasing compounds such as sulfur dioxide and nitrogen oxides into the air. These react with water to produce acidic chemicals that fall with rain. The evidence of its effects can be seen in the deterioration of many of our forests, and the damage is becoming more and more obvious.

Human activities can affect soil in other ways as well. For instance, poor farming practice can lead to a different type of nutrient leeching. Crops take up the nutrients as they grow, and then we pull the plants from the ground and ship them away. As a result, the nutrients that were normally recycled locally (soil to plant and back again), now get transported to your neighborhood's supermarket shelf. The next generation of plants doesn't grow as well because most of the Mg^{2+}, say, is now sitting in "aisle three" next to the cheesy poofs, a tofu turkey and a plastic dinosaur-shaped neon-green sandwich cutter. To battle this, modern farmers often replace the leeched nutrients by spreading fertilizers and animal dung on their fields. In the latter respect, I'm quite glad that 'organic' produce is available at my local grocer. After all, only the finest veneer of cow poo will do for me, thank you.

Pulling crops away from fields isn't the only way in which poor farming practice can wreck a local ecosystem. Sometimes, poorly planned irrigation can destroy a field as well. For instance, if standing fresh water is introduced on top of soil that is sitting above a lot of salt – such as in farmlands near sea level – the newly introduced fresh water can pull salt up from the ground. As the water evaporates, the salt begins to accumulate on the surface, radically changing the life-sustaining qualities of the field. There is some evidence to suggest that many of the fertile lands of Ancient Middle East have been destroyed by this kind of irrigation. The solution to this appears to be slower irrigation – less fresh water standing around – although this remains to be proven.

Lastly, removal of vegetation can speed nutrient leeching in a way unrelated to the direct removal of nutrients that go with the plants. The roots of the killed plants die, and as they die they lose their ability to hold the soil together. When this happens, water can flow away from the area more quickly, resulting in nutrient leeching and rapid erosion. If you don't like a hill, remove all the plants on it and watch what starts to happen.

That just about wraps up the first half of our look at biogeography. Let's have a short break and pick it up again in the next unit. See you then.

Summary

In this chapter, we learned about some basic concepts in biogeography.

- The Earth gets almost all its energy from the Sun. Sunlight heats the Earth, and much of this heat is lost to space. The remainder is trapped as potential energy in the form of new chemical bonds, altered weather, changes in the environment and so forth.
- The equatorial region of the Earth receives the most intense sunlight. This causes air and water to rise up. As the air rises, the water comes back down, resulting in lots of rain near the equator. The dry air moves north and south, dropping around the Tropics of Cancer and Capricorn. These latter areas tend to be very dry.
- Wind patterns are caused by a variety of phenomenon. Sea breezes occur because of differences in temperature between land and sea. During the day, the land is warmer than the sea, and rising land air drags cooler sea air toward shore. During the night, the warmer sea air rises, dragging cooler land air away from shore.
- Heated air tends to have higher pressure than cold air. All other things being equal, high pressure air moves toward low pressure air, in the form of winds.
- The Coriolis Effect causes wind to travel in circular patterns due to the rotation of the Earth. This happens because wind at one latitude has a different rotational speed than the Earth at a different latitude. So, when wind travels away from the equator, it deflects east. When it travels toward the equator, it deflects west. The combination of the two leads to circular wind, as is seen in hurricanes.

- The Moon influences wind by pulling air toward the side of the Earth closest to it. Tides result from the same kind of process, and water currents are similarly influenced by it as well.
- Mountains affect the direction of wind flow by providing a physical barrier, deflecting air flow or forcing air higher into the atmosphere. When wind tries to cross a mountain chain, the increase in altitude causes the air to cool, which forces it to drop its water on the near side of the mountain. The far side tends to have a dry environment as a result.
- Cold air rushing into a warm region will cause warm air to rapidly rise, leading to thunderstorms. This is called a cold front.
- Warm air rushing into a cold region will slowly rise up in altitude, leading to prolonged rains. This is called a warm front.
- The closer to the equator you go, the more photosynthesis tends to be done. This is because of the intensity of light in that region. As you progress toward the poles, less light is available, so the amount of photosynthesis drops off.
- All the energy captured by photosynthesis is called the gross primary productivity of the Earth. Part of this is used for respiration. The part that remains is the net primary productivity, which is the amount available to the remaining organisms. The bulk of life draws its energy from the net primary productivity.
- Most photosynthesis occurs in the rainforests and oceans. Next are the deciduous forests, boreal forests, savannas and grasslands. Lastly, deserts and the tundra do the least.
- Energy flowing through an ecosystem is usually described as a food chain. There are many food chains within an ecosystem, and all these together comprise the food web.

- At the very bottom of the food web are the primary producers – the autotrophs. These have the greatest amount of energy in the ecosystem. Next are the primary consumers – the herbivores – that live by consuming autotrophs. These individuals make up the second energy level (or second trophic level) of the ecosystem. Last are the carnivores, which live by eating the herbivores. This is the third (and sometimes fourth and fifth) trophic level. Collectively, all the trophic levels form the shape of a pyramid, where most energy is at the bottom and progressively less is at the top.
- The pyramid model can also be used to show the total biomass of each trophic level. The primary producers occupy the most mass, followed by the primary consumers, then secondary consumers and so forth. Some species occupy multiple levels of the pyramid model, depending on the different ways in which they acquire their energy.
- The decomposers are not represented in the pyramid model, because they feed on all organisms.
- The atoms and molecules used in the reactions of life must be continually recycled because the Earth is a closed system. Almost nothing gets in or out.
- Water cycles back and forth between the atmosphere and the ground via evaporation and rain. Water flows downhill into streams, rivers and lakes, eventually going to the ocean. Water absorbed by plants gets drawn up into leaves where it gets put into glucose molecules or "sweated" by plant transpiration. Water can also soak into the ground and find its way to oceans and lakes via underground routes or get stuck in underground lakes.
- Carbon is pulled from the air by photosynthesis. It is released by respiration. Most of the carbon in dead biological material is released by decomposers, such as worms and bacteria. The

carbon that is not freed gets stuck in deep layers of soil and eventually turns into coal and oil over geological time. Recently, humans have begun to release this trapped carbon through industrial combustion, and the rapid release has started to destabilize the atmosphere.

- Nitrogen is pulled from the air by nitrogen fixing bacteria using enzymes such as nitrogenase. The resulting nitrogen compounds are taken up by plants and animals. Eventually, these compounds are converted back into N_2 and released into the air by denitrifying bacteria. Recently, humans have begun pulling nitrogen from the air using industrial methods.
- The oxygen cycle starts with oxygen bound within water and carbon dioxide. These are incorporated into organic molecules during photosynthesis. In addition, some oxygen is released as O_2 during this process as well. This O_2 is then used during respiration to release water and carbon dioxide, completing the cycle.
- The phosphorus cycle starts with phosphorus in exposed rocks. Water erodes the rocks, releasing the mineral into streams, rivers and other waterways. Once dissolved in water, phosphorus is taken up by plants and animals. It is also deposited in the soil. As the soil gets deeper and deeper, the phosphorus gets compacted into newly forming rocks, and over geological periods it gets moved around by the large forces that work upon the planet. Eventually, the rocks get exposed again, starting the cycle over.
- Humans create a great deal of new compounds, few of which get fully tested for their effects on the environment.
    - Some compounds harm the environment directly. Other compounds are harmless until the environment changes them into a more harmful form.

    - Some compounds require great quantities in order to cause harm. Others do damage in small amounts.
    - Some compounds affect the physical environment, such as fluorocarbons vs. the ozone layer. Other compounds are poisonous to some species in the environment.
    - Chemicals can increase in concentration within a food chain via biological magnification. DDT is a good example and has resulted in the near extinction of the American Bald Eagle.
- Soil is mainly comprised of two parts: dead organic matter and inorganic particles.
    - Inorganic soil particles consist of silicon or aluminum compounds. The particles are divided arbitrarily into three sizes. The small particles make up clay. The medium particles make up silt, and the large particles make up sand. Clay tends to hold water too tightly. Sand tends to hold water too poorly. They are poor substrates for plant growth. Silt soils typically hold water much better and serve as better substrates for plants.
    - High water runoff can cause nutrient leeching by dragging away many of the minerals within the soil.
    - Acid rain can accentuate nutrient leeching by dislodging cations from inorganic particles. Examples include K^+, Mn^{2+}, Ca^{2+} and Fe^{2+}. Once dislodged, they can get washed away and become inaccessible to plants and animals.
    - Crop harvesting can leech nutrients from fields because the nutrients are carried away with the crops. To correct this, farmers frequently apply fertilizers to their fields.

- Standing water from irrigation can pull salt from the ground if the farmland is too close to sea level. The salt gathers when the water evaporates and eventually makes the soil no longer able to support plant growth.

Unit 5

Biogeography II

In this section we cover:

1) The Major Biomes of Earth
2) Continental Drift
3) Barriers to species migration

Land Biomes

The various physical parameters that exist around our planet – from temperature and rainfall, wind and humidity, latitude, mountains and soil quality – have given rise to a number of ecosystem templates, called biomes. There are a number of these biomes on our planet, some becoming less common and others more so. They include places like the **tundra**, **taiga**, **temperate deciduous forests**, **rainforests**, **grasslands**, **deserts** and coffee shops.[46] We'll discuss many of these here.

---

[46] Ok. I'm kidding about the coffee shops, but in all honesty, I wouldn't be surprised if biologists of the distant future included 'cities' as one of the Earth's biomes – or *only* biome, sadly. After all, humans are a result of natural evolution. That means we *are* natural, so shouldn't we include our behavior as an extension of nature? I mean, a beaver's dam radically changes its environment, but we consider it to be natural, right?

Tundra

The tundra is a circumglobal strip of area just south of the artic in northern Canada and Russia. It has a permanently frozen layer of subsoil. As a result, the tundra has a difficult time sustaining large plants and is occupied mainly by mosses, lichens, perennial plants and small grasses. The summers are warm enough to allow the flowering plants to bloom, transforming the region into a vibrant display of colors. In addition, these few warm months melt the very top layer of frozen soil, making the ground mushy. The tundra is dotted by lots of lakes, ponds, and marshes, and is host to a small number of critters. Some of the representative animals are arctic wolves, foxes and rabbit as well as reindeer and caribou.

Taiga (Boreal forest)

The taiga (or **boreal forest**) is the region south of the tundra. It too is ring shaped, extending around the globe. Unlike the tundra, its subsoil is not permanently frozen, thawing for parts of the year to allow more vegetation to take hold and grow. Conifer trees are among the most dominant vegetation there. The region also supports a larger number of animal species than the tundra, including moose, black bears, wolverines, porcupines, wolves and lynx. Also like the tundra, the taiga is dotted by numerous ponds, lakes, streams and marshes. And no, the taiga is not related to cats.

Deciduous Forests

The tundra and taiga are both circumglobal rings, centered on the poles. Their main characteristic is their latitudes; they are so close to the poles that their soil is frozen for at least part of the year. As you move further away from the poles, the soil becomes less frozen and more hospitable. This yields ecosystems capable of supporting much larger vegetation. In areas where moisture is readily available, deciduous forests grow. They have a lot of hardwood (broadleaf) trees, such as elm, birch, maple and oak, among countless others. Deciduous forests offer a large number of vertically stratified micro-ecosystems as well, which support a variety of smaller plants and animals. The total number of plant and animal species is much larger than the tundra and taiga, encroaching upon a size too large to list. The most represented are mammals, birds and insects. The least represented are the reptiles and amphibians, although a few can be found here and there. Common animals include foxes, bears, deer, squirrels, humans and other rodents.[47]

Grasslands and Shrublands

In temperate areas where the moisture is not sufficient to support forests, grasslands and shrublands develop. The grasslands consist of fields of herbs and grasses, with the occasional island of trees here and there. They are often characterized by warm-cold or wet-dry annual cycles, depending on the location. Because of the cold and/or dry parts of the year, large vegetation has a hard time taking hold. As a result, grasslands are less complex than most of the other biomes. They support a number of rodents and are often occupied by a few large herbivores. The grasslands of North America, for instance, used to be home to many herds of buffalo, and the grasslands of

[47] Humans are not considered rodents, of course, but at times I wonder. . .

Africa (also called savannas) are often the home of many zebra, gazelles and elephants.

The term **shrublands** (or **scrubland**) refers to an area where the dominant vegetation consists of bushes, scrubs and small trees, like a field of giant green and prickly cotton balls. Some of them receive higher levels of yearly rainfall than grasslands. Others receive less and can be found predominantly around deserts. The factors leading to grassland versus shrubland include rainfall, temperature, soil content and so forth, but as a first degree of approximation you can think of shrublands as being a midway point between grasslands and temperate forests.

Tropical Rainforest

Tropical rainforests are like deciduous forests taken to the next level. These are some of the most complex ecosystems on the planet and can be found in the most moist and warm land areas. By far, most are contained within the tropical zone, i.e. south of the Tropic of Cancer and north of the Tropic of Capricorn. Other tropical rainforests can be found on island climates that receive abundant warm winds and moistures, as is common in many Pacific regions. Tropical forests can grow to be several hundred feet tall, allowing incredibly complex and rich vertical micro-ecosystems to develop. At the very top of tropical forests, the thick tree canopy blocks practically all sunlight from the areas below, creating a dark world of underbrush full of highly efficient broadleaf vegetation and a plethora of almost every imaginable critter. Reptiles, amphibians, birds, insects, mammals and whip-wielding archeologists are all represented. In addition, the thick growth of vegetation stifles the wind, yielding an icky soup of humid air. Altogether, these biology-gone-wild ecosystems fix between 10-20% of the Earth's total carbon. Sadly, rainforests are rapidly disappearing, owing to a species of highly pervasive coffee-shop-creating bipeds that occupy the rest of the globe. In the place of jungles, we are sure to get an entirely different kind of biome, perhaps a parking lot or shopping mall. Keep your fingers crossed.

Deserts

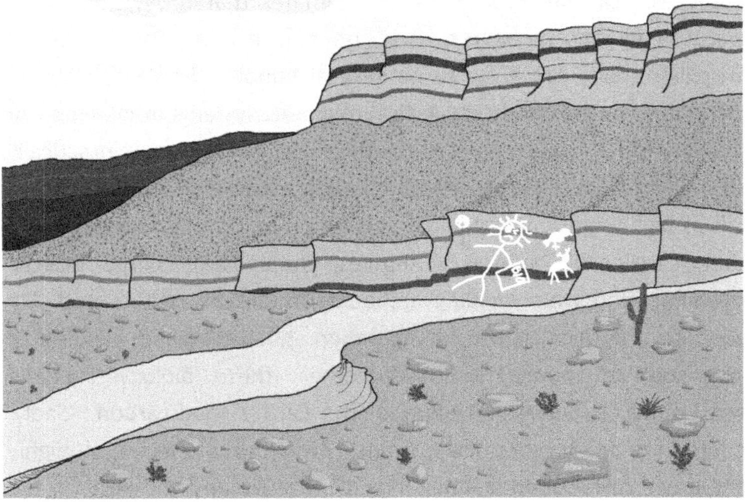

The last of the biomes that we will discuss are the deserts. Each is unique, but they all have one thing in common: low annual rainfall. As a result, they contain very little vegetation. Some deserts, such as the Sahara Desert, have so few vegetation that it is little more than sand dunes. Others, such as the Mohave Desert, have a variety of hardy, drought-resistant succulents, such as cacti. Still others have a brief wet period, during which fast growing annual plants germinate, produce seeds and die. The seeds lay dormant until the next season. Since deserts have very little vegetation, they have poor temperature moderation. The days are blistering hot and the nights sometimes quite frigid. Deserts support a number of different organisms, such as lizards, snakes, mice, coyotes, insects and various birds, but compared to other ecosystems, they are much less complex.

In the next image, we see the relative spread of the various biomes. Notice how the rainforests are largely contained within the strip between the two tropic lines and how many of the deserts fall roughly under the tropic lines. Further, notice how grasslands are often positioned between deserts and forests.

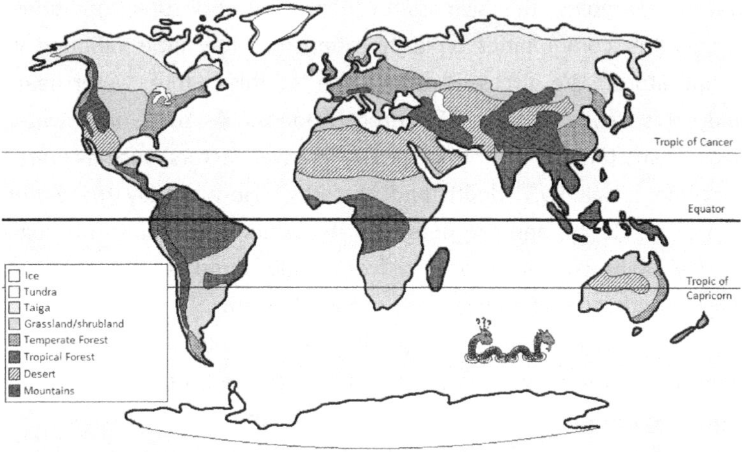

Effects of Altitude

Distance from the equator is one of the primary factors in determining the average temperature experienced by a location. Another important factor is altitude. The higher you go, the colder it gets. As a result, an increase in altitude often mimics the effect of increasing latitude. More explicitly, at the very tops of mountains you see regions that are equivalent to the Earth's icy poles. As you move down the mountain, you begin to see frozen tundra areas, then taiga, then finally a blend into whatever biome is at the bottom. If a mountain face is on the windward side, the water frequently dumped there will result in deciduous forests or rainforests at the bottom, (depending on the amount of rain and the temperature). On the opposite side, you'll typically see drier biomes, like grasslands or deserts. It all depends on the finite conditions.

Aquatic Ecosystems

Land biomes form largely as a result of moisture and temperature, but these two parameters are largely irrelevant under water. Forgoing the silliness of pondering moisture conditions, underwater communities typically experience very little variation in temperature. We discussed the reason for this before: water heats and cools slowly because of hydrogen bonds. As such, underwater ecosystems tend to form in response to other factors, such as water movement, salinity,[48] depth and so forth. How quickly the water moves, its depth, and the nature of its motion can result in a huge variety of unique environment, from shallow rapids, to deep and turgid ponds, to vast and dynamic oceanic bottoms.

When it comes to aquatic environments, we often divide them up in a variety of ways. For instance, the bottom part of the marine environment is called the **benthic zone**. This is where the bottom feeders are. The water above – and all things suspended in it – is considered part of the **pelagic zone**. When it comes to oceanic ecosystems, we sometimes divide them up based on how far they are from the shore. Those nearest to the shore in the areas pelted by the regular actions of the waves are in the **littoral zone**. Further out is the **neritic zone**, which extends to the drop off of the continental shelf. Ecosystems in the deep part of the ocean, beyond the drop off are in the **oceanic zone**.

Lastly, we can divide up aquatic ecosystems based on depth. The first 200 m is often referred to as the **photic zone**. "Photic" means 'light,' so the photic zone is the lighted part. The first 100 m in particular is called the **euphotic zone**, which means 'true light,' meaning that it's the brightest of the lighted part. Below 200 m is the **aphotic zone**, which just means it's the dark part.

---

[48] Salinity = how salty the water is.

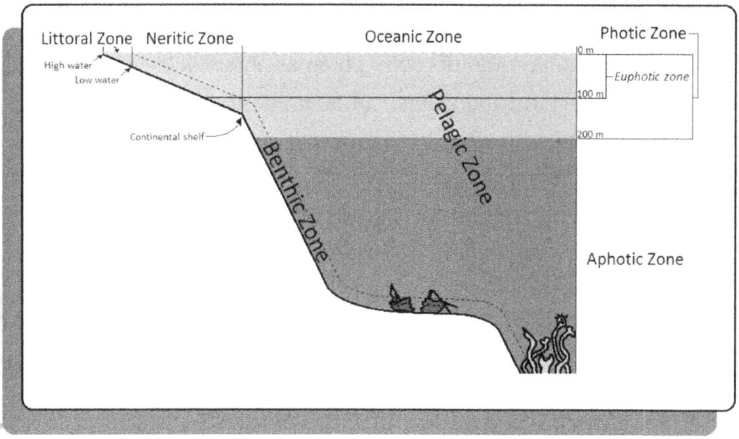

The bulk of the photosynthetic region of aquatic systems occurs mostly in the first meter of water. Most of the aquatic bottom communities (in the benthic zone) are comprised of organisms that 1) feed on floating protists, bacteria and other tiny critters, or 2) sift through dead materials that drift down from the photosynthetic layer above. The most grandiose of these ecosystems are the coral reefs, which are often referred to as the rainforests of the oceans. These colorful communities are created by **cnidarian** organisms that build their homes using calcium-based compounds. These form rocky but fragile structures upon which grow a huge variety of other organisms. The cnidarians feed off of zooplankton, floating nutrients and dead debris, although some also have symbiotic photosynthetic organisms called **zooxanthellae** living in or with them. Unfortunately, these ecosystems take a great deal of time to form but are easily destroyed. As a result, human activity has caused many of these underwater wonders to vanish.[49]

---

[49] You can be sure that if we ever find a way to build a parking lot on the ocean floor, you can kiss these ecosystems goodbye.

Primitive Earth

We've seen some of today's biomes and how they are related to the position of land and water, but how did the land get to where it is today? All evidence points to Earth being about 4.5 billion years old. As recently as 250 million years ago, all the contents were clumped together in one big continent called **Pangaea.**[50] This giant land mass began to break apart around 200 million years ago into two smaller land masses (both still bigger than any of today's continents). The two resulting continents were **Laurasia** in the north and **Gondwana** in the south. Laurasia eventually gave rise to North America, Europe and most of Asia. Gondwana in turn gave rise to Africa, South America, India, Australia and Antarctica. This, of course, is putting it simply. Let's have a little deeper look at the details.

Based on what we can tell, Gondwana first split into a number of pieces prior to yielding the modern day continents. One piece contained Africa and South America (still attached to each other), which retained a connection to Laurasia. India broke away and moved north as well, eventually bumping into the southern part of Laurasia. In the process, the India landmass pushed up the Himalayan Mountains.[51] South America then split from Africa and soon rejoined with the southern-most part of Gondwana that comprised Antarctica and Australia. These three continents were connected to each other before South America and Australia finally broke away, drifting isolated for a long time. Eventually South America collided with North America via the small land bridge where Central America currently is.

Eventually Laurasia divided as well. North America broke off from what was to become Europe and began drifting west. Before doing so, it existed for a while in connection with Africa. Granted, all

---

[50] What it looked like prior to that is not entirely known.
[51] Like what happens if two rugs are pushed against each other.

of this is probably as clear as mud to you, so I've included the image to the left to approximate what these continents looked like.

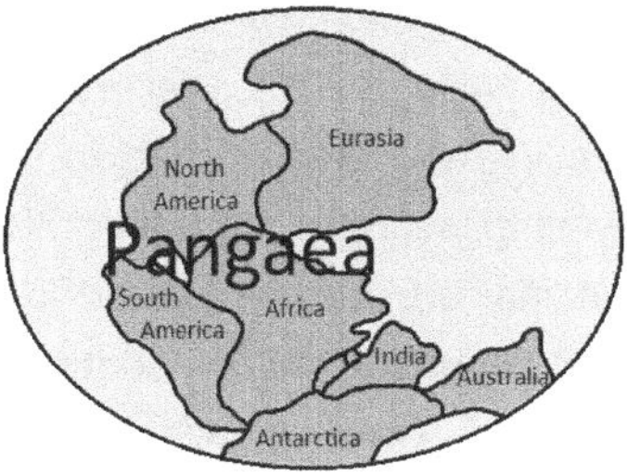

~250 million years ago

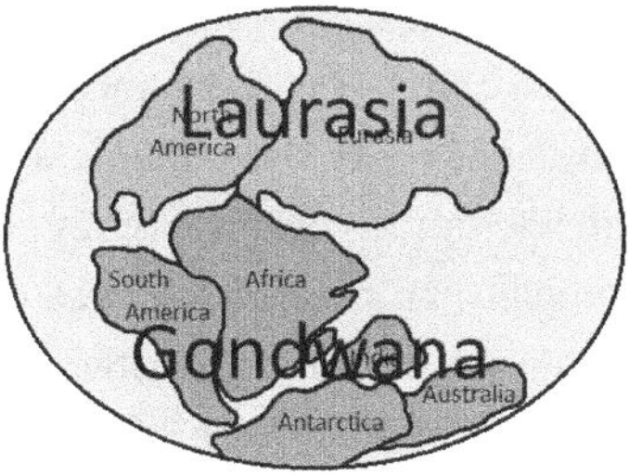

~200 million years ago

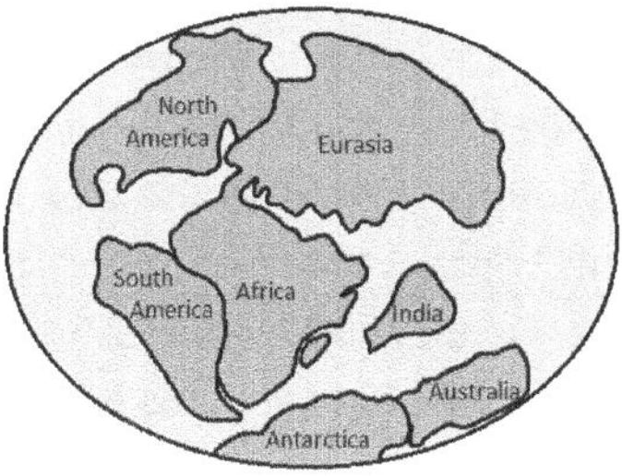

~150 million years ago

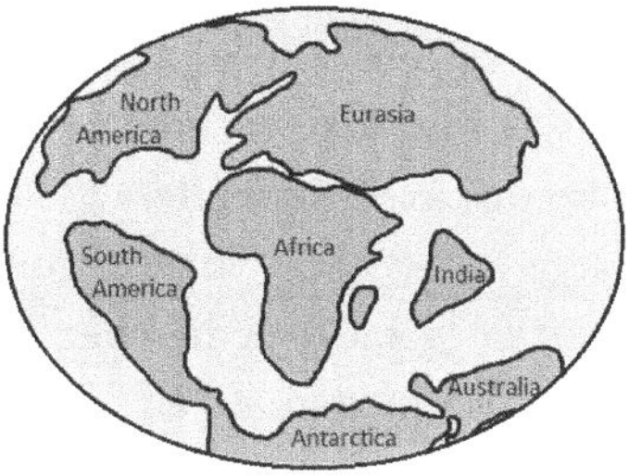

~50 million years ago

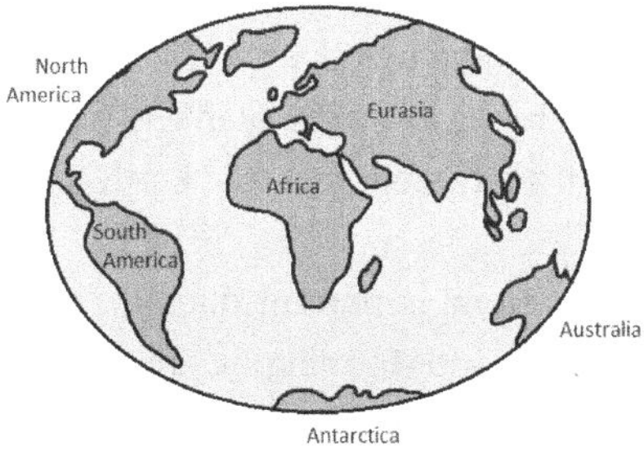

Present Day

So how do we know all of this? Well, the entire story is rather long and complex, but the pieces were put together based on a number of observations. First, the shapes of many of today's continents look like they fit together, the most obvious one being the facing borders of South America and Africa. This was the first clue. Second is the fossil record. For instance, large fossils found in both South America and Africa of the same species could only be possible if the two continents were once connected. Also, the fossil records for many continents show plants and animals from biomes that do not match with the current biomes occupying those areas. Antarctica, for example, has many fossil records that indicate it once existed in a much warmer climate. Lastly, geological deposits from one continent commonly match those of other continents. Once again, this could only be true if the two continents were once connected.

How could continents move apart like this? Well, the prominent theory to explain this is the **plate tectonic theory**, which basically says that the continents are floating around atop the liquid molten core of the Earth. Most of us are not used to the idea of floating rocks, so "floating continents" may sound absurd. The important point to remember, however, is that rocks are much less dense than molten metals. As the metals melt and sink toward the *very hot* center of the Earth, they push up the rocks, which end up floating on the surface. As such, the continents drift around like pieces of wood bumping into each other, only they move so slowly that it takes millions of years to notice the difference. Once in a while, we get a reminder of their slow activity in the form of earthquakes[52] and volcanos. Apart from these things, the Earth is rather silent on the matter.

At any rate, several important outcomes have resulted from the continents moving and forming as they have. First, those continents that were the most isolated, such as Australia, have had the greatest amount of evolutionary deviation. This is the reason for Australia's bizarre line-up of species found only there (excluding the

[52] Earthquakes are believed to occur as a result of two land masses scraping against each other.

animals capable of flying or swimming there). South America has also experienced quite a bit of time isolated from the others . . . after splitting off from Antarctica during its brief connection to it. During its isolation, it too evolved a number of unique species. On the other hand, South America and Australia have a large number of marsupials on them. Antarctica has fossils of marsupials. From this, we can conclude that marsupials evolved when these three continents were still connected. In the modern world, North America has a few marsupials as well. They arrived here by migration *after* South America became physically attached via Central America. Out of all the continental bodies, the chunk that had Europe, North America, most of Asia and (for a long time) Africa were connected for what appears to be the longest time out of any of the great land masses. As a result, many of the fossil records and modern species are similar across these three continents. Throughout that time, the only real barriers to migration were the developing Himalayas, the Mediterranean Ocean and the early Sahara Desert. Despite these barriers, many of the species are similar to each other.

Speciation across Barriers

Throughout evolutionary history, species have migrated from place to place. We saw an example of this when I mentioned how marsupials migrated into North America from South America. Here, we'll have a look at the general principles that describe the phenomenon of species migration. For simplicity, we'll break it down into three considerations.

Physical Access

Before a species can migrate into a new ecosystem, it must have physical access to the new region. Oceans, mountains, impassible land and so forth can keep one species from entering into a new home. On the other hand, a physical barrier for one species may not necessarily represent a barrier for another. A chain link fence might stop my neighbors' rotten kids from getting into my yard, but it doesn't stop a bird . . . or their annoying yapping dog that constantly crawls under it. An open field will not stop a caribou but it's sure to stop a fish.

Throughout evolutionary time, oceans have been perhaps the strongest of all barriers, resulting in large differences between the species of one continent and another. Still, bodies of water are not entirely impassable, especially to small animals. Naturally forming rafts made of logs and other refuse are not uncommon, usually after storms. Any small animal caught on them can drift for hundreds of miles before finding a new home elsewhere. Sometimes, a good storm can whip through an area and carry small animals such as insects away on the wind. This is commonly how insects will colonize small islands or distant lands. At other times, an animal can catch a ride on a more mobile organism, such as a bird, a fish, an herbivore or whatever, and can be deposited far from its original home. Bacteria, fungi, protists, bugs and countless others species can migrate this way. In fact, many plant species rely quite heavily on other organisms for their dispersal. They pack their seeds into fruit, the fruit is eaten, and the seeds survive in the digestive tract until it is dropped off with the rest of the feces. During this process, the animal that ate the fruit carries the seed to some distant location.

Physical barriers change with time, sometimes rapidly and sometimes imperceptibly slow. When they do so, they often force the species they bar to move with them. If the bed of a north-south river slowly drifts west over several hundreds of years, the species of organisms stuck on the west side will be forced to keep moving west. If an ocean between two continents grows wider, the species on either side have no choice but to get farther apart.

Anatomical/Physiological Compatibility

Even when a physical barrier is not present, speciation is not guaranteed to succeed. The migrating species must be physiologically adaptable to a new region. For example, suppose one species of tree frog exists within a region of rainforest that contains an abundant amount of a specific food type. Then, suppose that several members of the tree frog species attempt to migrate to the far end of the rainforest where the same food has a slightly higher level of a toxin. The presence of toxin may cause the frog to be unable to thrive in that location. As such, no physical barrier stopped them; they just weren't physiologically adapted to the types of food found in the far end of the forest.

Physiological compatibility extends to much more than just food. There may be differences in temperature, moisture, light and so forth in the new region. Even more importantly, the new region may have different predators or other species of tree frog that compete fiercely with the newcomers. Sometimes the new environment might be perfect for the new species, except for sporadic frosts that come through. Even a brief weather abnormality might be enough to make a new environment impossible to live in. It would be like human astronauts finding an Earth-like planet that is perfect for habitation, except for the one day out of the year when the planet heats up to 3,000 degrees Celsius. Such a temporary *inconvenience* would stop our attempts to colonize the new planet . . . or in the very least make it exceptionally hard.

As you hopefully can see, any number of factors may prevent the spread of a species from one region to another, and collectively these restrictions are known as the **Law of Tolerance**, which basically just says that the species will be limited by its ability to tolerate subtle differences between the original and new niches.

Other Environmental Factors

There are a number of other environmental factors limiting the spread of species as well. One is the size of the new environment. For instance, small environments often carry smaller odds of colonization success than larger regions. The reason is arguably due to the limitation of resources. Consider a bunch of spiders blown away by a storm to land on a distant island. Let's take two cases. In the first case, let's assume the island is only thirty square meters in area. In the second case, let's say the island is thirty *thousand* square meters. The islands are otherwise identical. The spiders landing on the first island are less likely to survive because they have access to fewer resources.

Another ecological factor in species migration is complexity of the new environment. The more complex an environment is the greater are the odds of finding hiding places, nesting sites, novel food sources and other materials you might need to survive. Consider spiders landing on an island that is perfectly flat with little more than grass on it. They will be less likely to survive than similar spiders landing on an island of the same size occupied by a mountain and a rainforest.

The last ecological factor that is important in species migrate is the distance between the original ecosystem and the new ecosystem. The farther it is, the less likely the species will be able to get a foothold. Conversely, the closer the new environment is, the more members of your species will invade . . . as will members of *other* species as well. This aspect of speciation is often studied by examining islands close to shore. What we often find is that the rate of appearance of new species on such islands (or **rate of speciation**) depends heavily on the distance from the island to shore. Islands near the shore have a high rate of new species appearing during any given year. The farther away the island is the less new species will appear in the same given year.

On the other hand, the more species arrive on the island, the higher the competition and – consequently – the higher the **rate of extinction**. This latter aspect is heavily dependent on the size of the island. Small islands are battlegrounds with high rates of extinction. Large islands are more stable, with less species dying out during any given year.

Tying these two principles together, we find that at any given point in time, the total number of species that an island can support tends to equilibrate at a number in which the rate of speciation equals the rate of extinction.

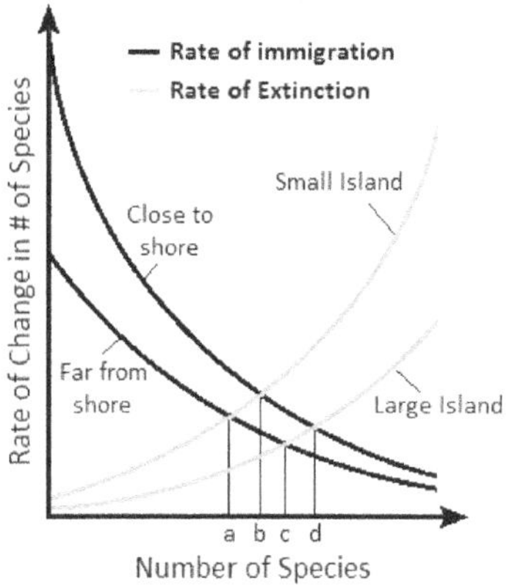

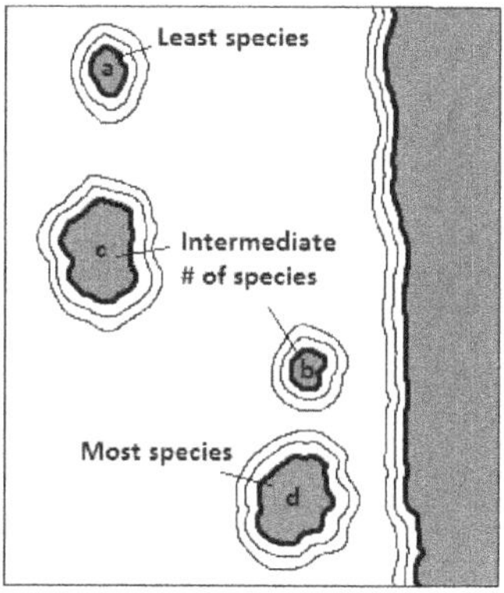

So an island that is, say, one mile offshore might be able to support 100 species because the rate of species migration is large, whereas the same island 50 miles offshore might only be able to support 30 species. In the latter case, the rate of new incoming species is far less, but so is the rate of extinction.

Summary

In this chapter, we learned about some of the basic biomes, the formation of today's continents and the basics of species migration.

- A biome is an ecological template. The major ones include: the tundra, the taiga, grasslands, deserts, deciduous forests and tropical rainforests.
- The tundra is a circular band-like region just south of the North Pole. It has permanently frozen soil and contains small plants and a few animals.
- The taiga is south of the tundra and is frozen for a large part of the year. It is dominated by coniferous trees. It contains wolves, moose and a variety of small animals and birds. Its other name is the boreal forest.
- Grasslands are regions that are warm enough to support forests but receive insufficient rainfall to grow them. They are occupied by grasses, small shrubs and pockets of trees. They are home to small rodents and herds of grazing herbivores.
- Deciduous forests are ecological regions that are sufficiently warm and moist enough to house a permanent collection of deciduous (leaf-shedding) trees and a small amount of undergrowth. They contain vertical stratification and have greater numbers of species than grasslands.
- Tropical rainforests have rich collections of dense vegetation and towering trees, with dark and moist undergrowth. They are primarily found within the tropical belt and have a rich number of species from practically all the major animal groups. The tropical rainforests are rapidly disappearing due to deforestation by humans.

S

T

V

W

Z

Made in the USA
Monee, IL
02 May 2026

49359333R00066